Strange Science

Strange Science

Investigating the Limits of Knowledge in the Victorian Age

LARA KARPENKO
AND
SHALYN CLAGGETT
EDITORS

University of Michigan Press
Ann Arbor

Published in the United States of America by the
University of Michigan Press
Manufactured in the United States of America
♾ Printed on acid-free paper

2020 2019 2018 2017 4 3 2 1

A CIP catalog record for this book is available from the British Library.

Library of Congress Cataloging-in-Publication Data

Names: Karpenko, Lara Pauline, editor. | Claggett, Shalyn R., editor.
Title: Strange science : investigating the limits of knowledge in the Victorian Age / Lara Karpenko and Shalyn Claggett, editors.
Description: Ann Arbor : University of Michigan Press, [2017] | Includes bibliographical references and index.
Identifiers: LCCN 2016035928| ISBN 9780472130177 (hardcover : alk. paper) | ISBN 9780472122455 (e-book)
Subjects: LCSH: Science—Social aspects—Great Britain—History—19th century. | Science—Great Britain—History—19th century. | Great Britain—History—Victoria, 1837–1901. | Parapsychology—Great Britain—History—19th century.
Classification: LCC Q127.G5 S77 2017 | DDC 509.41/09034—dc23
LC record available at https://lccn.loc.gov/2016035928

Foreword

Dame Gillian Beer

In this volume, *Strange Science*, the editors Lara Karpenko and Shalyn Claggett emphasize the borders of investigation in their subtitle, *Investigating the Limits of Knowledge in the Victorian Age.* "Limits" here are subjects for fresh investigation rather than clamping containers. Beyond the current limits are thriving new territories, or delusive dream countries. This powerful collection gathers examples of both. But a number of the essays also make it clear that even work that fails or where the procedures are chaotic and the assumptions doubtful may eventually find some presence in later discoveries. Work that challenges dominant assumptions may provoke different kinds of insight from more orthodox workers over time.

Science is preoccupied with discovery and with justification. To that degree, all science seeks the strange. When found, the aim is to find a place for the discovery within the known system or, more radically and more rarely, to change the system. The struggle between novelty and affirmation of the known gives the zest to much scientific work. It demands cautious procedures and audacious guesses at the same time. Innovation and repetition are both essential. That much can be said of scientific work across fields and across time. But there are major differences between the practices of knowledge-seekers in the nineteenth century and the present day. One difference is the emphasis then on individual investigation rather than teamwork and the somewhat belated arrival later in the century of university laboratories with an array of instrumentation.

The essays here are concentrated in Britain so that the research cultures described are those of people in Victoria's reign, living within arguments and assumptions about empire, gender, and class that may—or may not—be unfamiliar now. Most of the people discussed in this book could assume a postal system that in London made deliveries at least ten times a day; they took for granted prodigious letter writing and intricate face-to-face contacts within social groups. Even within this kingdom diverse methods and enquiries were being pursued. The sheer variety of research cultures in different parts of the British Isles during the nineteenth century has been explored in the collection of essays edited by David N. Livingstone and W. J. Withers, *Geographies of Nineteenth-Century Science.*[1]

Empire produced some hierarchical delusions but it also propelled British people across the world and gave many an intimate familiarity with places remote from these islands. The passion for collecting, whether birds' eggs or butterflies, stones or big game, seems to have been unhampered by qualms about its effects. It ravaged some species; it also allowed an exquisite awareness of minute differences from example to example. It fueled taxonomic sophistication and it gratified the urge to possess, which is always a tempered or intemperate element in knowledge-gathering. The rich array of general journals, many of which discussed scientific matters alongside political, literary, and local issues, meant that the sciences were present in ordinary conversation among the educated. Geoffrey Cantor and Sally Shuttleworth demonstrated that in their edited collection *Science Serialized: Representations of the Sciences in Nineteenth-Century Periodicals* and in their website *Science in the Nineteenth-Century Periodical.* Shuttleworth is now leading a major investigation, "Constructing Scientific Communities: Citizen Science in the Nineteenth and Twenty-first Centuries," which is producing fresh knowledge about the contribution of amateurs to scientific projects in the Victorian era as well as in our own.[2] There is room for much more work on the contributions of Victorian working people to scientific knowledge-gathering.

How, then, to distinguish between scientific work familiar and strange, orthodox and odd? As the collection of essays makes clear, some enquiries that may now seem strange were, for a time at least, accepted as potentially mainstream science. The most famous of these is spiritualism, with the careful, even skeptical, evidence-gathering of the British Society for Psychical Research drawing in important scientists such as the distinguished chemist and physicist Sir William Crookes. The Society's *Phantasms of the Living*, as L. Anne Delgado comments, shows "the unsteady nature of both human perception and the knowledge that perception itself produces." Nevertheless, their use of witness statements chimes with current interest in individual accounts of phenomena, and their concern with perceptual bias even connects forward to some of the preoccupations of neuroscience now. And mesmerism, as Karpenko's essay suggests, has never gone away.

Strange science has a way of leaving traces for later workers to pursue. As Barri Gold points out, the continuity between traditional and more exotic theories relies on "the principle . . . that what we learn or hypothesize about the natural world must be consistent with what we already know," which "constrains and shapes any paradigm shift within the sciences." Yet what we know about the natural world is itself shifting; indeed, exploring and opening up those shifts is a fundamental and central concern of scientific enquiry. It becomes clear that no easy and permanent boundaries exist between the sober known and the extreme imagination. W. K. Clifford, whose geometric Clifford algebras now underpin many advances in physics and computing (though they were beyond the capacity of most of his contemporaries), asked his contemporaries in the 1870s to question even the uniformity of nature and to be skeptical of induction since there may always be another case not yet known or encompassed by the definitive current laws. Clifford was married to the novelist Lucy Lane Clifford and himself enjoyed writing fairy stories. His central demand was for the constant testing of belief by evidence. This demand does not exclude imagination. Like John Tyndall, another foresighted thinker whose work is alluded to in a number of the essays here, Clifford emphasizes the uses of imagining. He writes: "The scientific discovery appears first as the hypothesis of an analogy; and science tends to become independent of the hypothesis."[3] Before science can detach itself from the hypothetical, it must work through guesses and analogy. Though hypothesis may be superseded, analogy persists as a tool for understanding similarities, and for measuring differences.

So in Victorian scientific writing, as the contributors to this collection often brilliantly show, disciplines, systems, boundaries, fields, and the constraints of gatekeeping lie alongside overlaps, leakages, struggles, analogies, and fault lines. All these terms appear in the current collection, and together they aptly suggest the degree of reciprocity between way-out and conventional thinking in the period. The editors have organized the volume under the topics of plants, bodies, and energies, and the three parts address different subjects of scientific enquiry. Their concerns are coherent with each other.

One of the most fascinating outcomes of collecting together these essays by diverse hands is that certain works emerge strongly in different essays and begin to suggest fresh patterns for understanding scientific controversies of the period. One such work is Balfour Stewart and Peter Guthrie Tait's *The Unseen Universe, or Physical Speculations on a Future State* (1875). This popular and notorious work (denounced by materialists such as Tyndall and Clifford) was authored by well-respected physicists. Indeed, Tait was the close friend and major correspondent of the great mathematical physicist James Clerk Maxwell. Their argument attempts to recuperate energy and to point beyond entropy. They use rhetoric and evidence and analogies between systems of very different scale to argue for an eventually Christian universe. Strange science, it's clear, was produced by orthodox scientists as well as by intelligent and trained outsiders such as Annie Besant, who repudiated many of the methods of scientific proof, and whose insights fueled thinking well past her lifetime, as Sumangala Bhattacharya here argues.

The relatively slight presence of women workers among the topics of essays here (Marianne North and Annie Besant being honorable exceptions) is of course symptomatic of the exclusion of women from universities, societies, and public scientific laboratories in the Victorian period. When the British Association for the Advancement of Science decided not to admit women as full members, they nevertheless did open their lectures to both sexes, and women were an important part of their audiences. Popularizing was considered suitable for women writers, and perhaps partly for that reason their work would not figure in the outer reaches of enquiry but would stay close to orthodox science. Recent investigation has reminded us of women whose research was both central and innovative, such as the mathematician and electrical scientist Hertha Ayrton, even if their social position in relation to other scientists was peripheral.

A particular gain of the collection is the emphasis on forgotten aspects of well-known scientists' work. Francis Galton, famous as statistician and eugenicist, here emerges in Danielle Coriale's essay as a pioneer in the study of deafness and "the auditory imagination." The Bell family, well before the phonograph and the systemization of phonetics, were involved in extraordinary demonstrations of "Visible Speech." James Emmott's essay describes in lucid detail "how the fields of physiology, phonetics, and phonography are mutually determined in the

1860s and 1870s, and how each draws on a shared understanding of articulatory performance." Read the essay to discover why this is such a thrilling insight.

Respect for unexpected or unconsidered sources of experience and intelligence fuels the volume. This is very striking in the first part, on plant life and Victorian excitement about the senses and the erotic life of plants. Plant intelligence is the special theme of these essays, and as Elizabeth Chang observes: "To imagine how the organic world imagines the human is also to confront the limits of the possibilities of imagination." Lying behind that fascination, and that tonic realization of human limits, is the work of Erasmus Darwin in his poem *The Loves of the Plants* (1790) and his scholarly prose work *Zoonomia* (1794–96), which developed the theory of a common ancestor behind all living things and emphasized variability and evolution as a precursor to his grandson Charles Darwin. Even more, Erasmus Darwin developed an understanding of the sense-life of plants and its capacity to cast light on human senses. He was much mocked among his contemporaries. His work now stands as an example of strange science become familiar, and yet still with the capacity to surprise. That resilience is shared by a number of the figures discussed in this stimulating volume. Failure is not shameful when the work has been ardently pursued and the whole collection reminds us that enquiries that run askance the current norm may yet open unforeseen pathways for future workers.

Notes

1. David N. Livingstone and W. J. Withers, eds., *Geographies of Nineteenth-Century Science* (Chicago: University of Chicago Press, 2011). See also David N. Livingstone, *Putting Science in Its Place: Geographies of Scientific Knowledge* (Chicago: University of Chicago Press, 2013).

2. Geoffrey Cantor and Sally Shuttleworth, *Science Serialized: Representations of the Sciences in Nineteenth-Century Periodicals* (Cambridge: MIT Press, 2004); Geoffrey Cantor and Sally Shuttleworth, *Science in the Nineteenth-Century Periodical*, December 18, 2007, http://www.sciper.org/index.html; Sally Shuttleworth, *Citizen Science in the 19th and 21st Centuries*, http://conscicom.org.

3. William Kingdon Clifford, "Conditions of Mental Development," *Lectures and Essays*, vol. 1 (London: Macmillan, 1901), 92.

Acknowledgments

We would like to begin by thanking our wonderful contributors for their exceptional scholarship, intellectual generosity, and unfailing good humor. They have all made this process a true pleasure throughout. We are also grateful to the editorial staff at the University of Michigan Press and the anonymous external readers for their helpful and detailed comments. Their insight and assistance has been instrumental in helping us revise the collection. Sincere thanks are due to Molly Walsh and Brittany Larson, who helped us to prepare the manuscript for publication. We would also be remiss if we did not thank Aaron McCollough for his enthusiastic support of the project in its early stages.

We would finally like to thank our families and friends for their encouragement throughout this process. Lara Karpenko thanks her departmental chair, Deirdre Keenan, for her tireless support of her colleagues and for her boundless enthusiasm. Her mentorship and friendship is deeply appreciated. Lara would also like to thank her mother, Christine Karpenko, for instilling an early love of Victorian literature, her father, Leonard Karpenko, for showing her the wonders of science through a telescope, and her husband, William Phelps, for his unfailing love and support. Shalyn Claggett thanks her parents, Sam and Sherrie Claggett; her husband, Matt Lavine; and her best friend, Lucy Canessa. She sincerely appreciates their continual encouragement and moral support in this and all things.

Contents

Strange Science

Introduction

Lara Karpenko and Shalyn Claggett

In an 1872 letter to the *London Times*, Henry Dircks, renowned engineer and self-described "worker in the sciences," gently chastises the editor for suggesting "that the time has arrived when scientific men should examine . . . Spiritualism."[1] Confidently proclaiming that "*no* really scientific man believes in spiritualism," Dircks derides the practice as a mere "pseudo-science" and seems to exclude almost any experimental or innovative practice from valid scientific endeavor.[2] But by the end of this letter that began with so much bravado, Dircks surprisingly speculates that one-tenth of spiritualist séances may, in fact, reveal scientific truths and rhapsodizes that "science is beset with . . . great wonders."[3] Far from drawing hard disciplinary boundaries around scientific practice, Dircks ultimately advances a notion of science that may be consistent and factual but is also aesthetic, poetic, and as magical as it is mathematical. To some extent, Dircks's career showcases the same disciplinary fluidity as his letter to the *Times*. Though his name may now be forgotten, Dircks's scientific work can still be seen today at places like Disneyland's "Haunted Mansion Ride," which features the translucent stage apparition popularly known as "Pepper's Ghost."[4] Created through manipulations of lighting and mirrored surfaces, "Pepper's Ghost" was at once the product of scientific advancement and the subject of popular spectacle. Though Dircks proudly defined himself as a "worker in the sciences," the very fact that he dedicated his talents to creating a ghostly mirage suggests he was little interested in a scientific career narrowly conceived. Dircks's stage career and his letter to the *Times* both indicate that for Victorian audiences, thinkers, and scientists, the category of the scientific—even for those who claimed otherwise—was remarkably if not jubilantly unstable and existed in a disorderly space marked by heterodox methods of inquiry.

It is within this disorderly space that we situate this volume. The essays in *Strange Science* investigate the epistemological and aesthetic imaginings that occurred in the hazy region between what we would now term "legitimate" and "illegitimate" scientific practice in nineteenth-century Britain. By examining these strange subjects and modes of inquiry, taking them not as foolhardy moments epitomizing the ignorance of a bygone era, but as serious investigations at the limits of knowledge, these essays offer fresh and inventive readings of sciences, texts, and practices that have often been overlooked or forgotten. *Strange Science* reveals, to use Mary Poovey's phrase, a "messy history" of scientific practice more fluidly defined, and the volume builds on Poovey's examination of what she terms the "modern fact."[5] As Poovey traces the origins of this abstract yet everyday "epistemological unit," her work puts pressure on the notion that "numbers . . . guarantee value-free description," suggesting instead that even our "most commonplace ideas" have a history (xxv, xiv). The essays in this volume similarly complicate the notion that objectivity, falsifiability, numerical analysis, and scientific endeavor are somehow natural partners. Poovey ends her study in the 1830s when, as she argues, facts effectively came to signal "theory-free representation" (xxv). This volume explores the chronological moment immediately following that of *The History of the Modern Fact,* investigating an elastic cultural moment: one in which the "fact" existed as a "value neutral and context-independent" concept, but also one in which the "fact" so conceived did not necessarily define scientific pursuits.[6] While the essays provide a posthistory to Poovey's study, they also provide a prehistory to Mel Y. Chen's twentieth- and twenty-first-century-focused investigation of what she intriguingly refers to as "animacy"—or the "richly affective territory of mediation between life and death, positivity and negativity, impulse and substance."[7] For Chen, animacy is a "rangy, somewhat unruly construct" that "trouble[s] and undo[es] stubborn binary systems of difference, including dynamism/stasis, life/death, subject/object, speech/nonspeech, human/animal, natural body/cyborg."[8] The essays in *Strange Science* offer explorations of similarly "rangy" and "unruly" subjects, such as atomic theory, entropy, and the afterlife, ultimately gesturing toward a marked fluidity in Victorian conceptions of the relations between human and nonhuman, life and nonlife, and the normative and the nonnormative. In doing so, *Strange Science* places a wide variety of seemingly disparate fields in conversation with one another. Though it may seem surprising (if not strange) that we group botany, mesmerism, physics, yogic practice, and psychophysical research in one volume, these essays suggest that that these disciplines and practices were often informed by a similar cultural logic—one that embraced modes of inquiry that were intuitive, extrasensory, and fueled by a common desire to subject the most fantastic flights of fancy to the hard scrutiny of science.

We use the phrase "strange science" to characterize the diverse subjects addressed in this volume in order to evoke a sense of the curious felt from two different perspectives: that which was strange for the Victorians, and that which may seem strange to us. For the Victorians, the phrase "strange science" was frequently used to describe the astonishment and awe that the possibilities of science inspired. To list just two examples: a reviewer for *All The Year Round* termed chemistry "that strange science which changes and transmutes substances," and the physician Charles Neidhard suggested that homeopathy was a "wonderful and strange science" that could unravel "the laws of nature."[9] As Neidhard's almost rapturous usage indicates, far from being an adjective of derision, the term "strange" was often invoked to express a feeling of wonder and an embrace of liminality. For members of a twenty-first-century audience, however, the scientific subjects discussed in this volume may seem "strange" because the nature of the inquiry is so unlike the highly professionalized science of today. Since the nineteenth century, the increasingly specialized language of science, the removed and highly controlled space of the laboratory, and the technological sophistication of instruments have contributed to making the work of science seem removed from the social world. Although the field of science studies has emerged to bridge this gap, it has paradoxically raised the stakes of the division by seeming to deny the reality of the external world which scientists attempt to establish and explain. Yet, as Bruno Latour argues in a section of *Pandora's Hope* titled "The Strange Invention of an 'Outside' World," it is the assumed division between a scientifically accessible objective reality and a separate social world that is truly "strange": "We do not need a social world to break the back of objective reality, nor an objective reality to silence the mob. It is quite simple, even though it may sound incredible in these times of the science wars: we are *not* at war."[10]

Even if this tension between the ahistorical, objective world of science and the social, subjective world of historical reality is an illusion, the division is nevertheless one we tend to assume—but one that the Victorian thinkers discussed in this volume did not.

Strange Science examines the unconventional, fringe, and heterodox in Victorian scientific inquiry at the levels of both methodology and subject. As these essays show, heterodox methods were often employed in conventional scientific fields, and, conversely, rigorously scientific approaches were adopted to study what, by today's standards, were completely unscientific subjects. In investigating this history, *Strange Science* conceptualizes "boundary lines" in ways that are radically different from the professional and political boundaries that characterize recent scholarship in science studies. Groundbreaking work by such scholars as Stephen Shapin, Roy Wallis, and Roger Cooter has demonstrated that conflict over what constitutes scientific legitimacy is frequently motivated by the specific political and social interests of the intellectual groups involved.[11] This dynamic, in fact, is so pervasive it has been formalized by the sociologist Thomas Gieryn as "boundary-work," the "attribution of selected characteristic to the institution of science . . . or purposes of constructing a social boundary that distinguishes some intellectual activities as 'non-science.'"[12] While such discussions have offered an important contribution to science studies, this approach has become so prevalent that it has become a kind of conceit, typified by scholarly titles that include the words "boundary" and "margin."[13] This tendency to scrutinize the way in which professional boundary lines are drawn focuses on sites of institutional and political contest and conflict. Such broader sociological and historical studies, however, necessarily exclude the speculative practice of Victorian scientific thinkers on either side of these emerging boundary lines, for whom the issues of legitimacy and authority were of far less importance than individual interests fueled by the possibilities science inspired.[14] Admittedly, there have been discrete histories of Victorian pseudosciences, such as Alison Winter's *Mesmerized* and Roger Luckhurst's *The Invention of Telepathy*, which focus on a single science from its popular emergence to the point of its near extinction. *Strange Science*, however, places a range of unconventional forms Victorian science in dialogue with one another and thus foregrounds how seemingly unrelated approaches were often informed by a similar belief in science's ability to reach beyond the apparent limits of empirical knowledge.[15]

By reorienting the perspective on the Victorian scientific context into one of possibility rather than conflict, these essays reveal that the relationship between orthodox and heterodox science is often far more complex than any binary distinction might suggest. Alison Winter has argued that often seemingly controversial forms of scientific practice were supported by a surprisingly large portion of the scientific establishment. As Winter explains, "fine-grained social histories" sometimes reveal that it was "often impossible for Victorians to agree on what counted as illicit or pseudoscience or medial quackery in specific instances," and further, that "within those undefined areas, researchers used their scientific work itself to develop the basic principles that would underpin that practice."[16] The essays in this volume employ just such a "fine-grained" approach and show that many Victorian thinkers embraced, and were even inspired by, the uncertainty and instability that characterized heterogeneous forms of scientific inquiry. Certainly, legitimacy was an issue of central concern to many, but it was not the only, or even the most important, concern for individual scientific thinkers. Focusing exclusively on the issue of credibility comes at the expense of excising the profound sense of wonder that pervaded an age of rapid scientific advancement and discovery. Rather than science at the "border" or "margin," the subjects of interest in these essays are situated at the limits of knowledge—at the very outer bounds of what seemed possible, and possible to be understood, in the natural world. For these men and women, the "boundary-work" of central importance was not intraprofessional, but ontological and even metaphysical, and they embraced many unusual and unconventional methods to enlarge the territory of the known universe.

Examining such scientific subjects and methodologies in the Victorian period is particularly important because it was at this time that the idea of modern science, as we now conceive it, began to take shape. In Britain, the most historically crystalizing moment for "science" understood as the study of the laws of nature (rather than referring to any form of systematic knowledge) was the first meeting of the British Association for the Advancement of Science (BAAS) in 1831. In fact, the word "scientist" did not even exist until 1834, when it was coined by William Whewell in response to Samuel Taylor Coleridge's objection to members of the BAAS referring to themselves as philosophers.[17] Even within the BAAS (to say nothing of the constellation of other scientific societies that preceded or developed in opposition to it), the inclusion, exclusion, and reshuffling of fields was a matter of constant discussion: phrenology was fiercely debated, "tideology" was enthusiastically embraced, and medicine was begrudgingly accepted and then marginalized.[18] As these debates suggest, the very idea of science and the constitution of its disciplinary categories were ambiguous at this time. By examining instances of scientific thinking that straddled emergent definitions and boundaries, this collection brings to light the exciting interconnections such instability made possible.

Even though the move toward formalizing the disciplinary distinction of the sciences began to take hold in the 1830s, the Victorian period was nevertheless a time of free and open exchange between science and culture. This was due in part to the accessible language used by scientific writers, whose work was being published in popular periodicals beside essays on politics, philosophy, and literature.[19] As Jay Clayton has argued, the intellectual landscape of the early nineteenth century was characterized by an "undisciplined culture" precisely because the sciences had yet to establish themselves as discrete disciplines, with such features as dedicated journals, conferences, and credentialing practices.[20] Similarly, Gillian Beer has offered the metaphor of "open fields" to describe the fluid nature of epistemological exchange between science and culture, and Laura Otis has pointed out that Victorian scientists established their credibility throughout the Victorian period through frequent references to the Western literary canon, which secured their authority as educated and cultured intellectuals.[21] "Open," "undisciplined," and dynamic, this period was marked by the interpenetration of literary, religious, scientific, philosophical, and artistic ideas, and this volume aims to demonstrate the extraordinary degree to which this exchange generated new speculations about the nature of the physical world.

As we point to the disciplinary and procedural fluidity at the heart of Victorian scientific practice, we simultaneously suggest that the objects of Victorian scientific inquiry themselves—plants, bodies, energies—often lacked stable definitions. Though Cary Wolfe rightly points to Donna Haraway's 1985 "Cyborg Manifesto" as "probably the locus classicus" of twentieth-century critical conceptions of the cyborg, our volume suggests that Victorian science, with its interest in matters such as visible speech or stimulated physiology, also laid the groundwork for expanding the limits of what we consider a human body; in other words, Victorian scientists anticipated, if not helped to invent, the notion of the "posthuman."[22] In this sense, our collection engages with Deborah Denenholz Morse and Martin Danahay's *Animals Dreams*, which in part explores the "subversion of the possibility of a human/animal distinction."[23] Whereas *Animal Dreams* focuses on Victorian interactions with animals—beings that we all define as sentient—*Strange Science* enlarges upon this discussion in order to illuminate Victorian conceptions of entities that we most often define as *non*sentient. As our opening section on botany suggests, plants loomed large in the Victorian scientific imagination and were often endowed with moralizing, erotic, or murderous capacities; so even discourse that is radically posthuman, such as Forianne Koechlin's "The Dignity of Plants" (a passionate plea for "the moral consideration of plants for their own sake") arguably found its genesis in Victorian thought.[24]

Of course, this volume is not arguing that Victorian thinkers and scientists participated in a radically postmodern project of eradicating all categories of identification or classification. Certainly, as science became increasingly professionalized, it often helped consolidate Western, masculine, upper- and middle-class agendas, as the common phrase "gentlemen of science" suggests. Critics like Jim Endersby and Alison Winter have examined how the professionalization and masculinization of the sciences often gave rise to intense cultural debates about which scientific subjects were appropriate for university study or for middle-class men to consider suitable as a profession: *Strange Science* participates in this rich critical conversation and invites readers to consider heterodox scientific inquiry in light of these gender constructions.[25] Along these lines, our collection also extends upon Evelyn Fox Keller and Helen E. Longino's landmark volume *Feminism and Science*, which argues that "scientific research and science-based technologies [participated] in the continuing subordination of women."[26] Following this observation, essays in our volume explore the tension between orthodox masculinist scientific discourses and feminized heterodox scientific investigation, suggesting that fringe sciences often provided women with an opportunity to create new knowledge within a field that would have otherwise been closed to them. As the essays in *Strange Science* examine the interplay between nonnormative scientific endeavors and gender construction, they also (sometimes simultaneously) comment upon Victorian discourses of empire and nationalism. Our investigation of these fringe and heterodox scientific endeavors also reveals that both actual and imagined contact zones (to use the language of Mary Louise Pratt) could sometimes be located in invisible, delicate, or surprising spaces: inside the atom, for example, atop the petals of an orchid, or deep within the intimate recesses of the human body.[27]

The collection is organized into three parts, each addressing a different object of scientific inquiry: plants, bodies, and energies. The essays in *Strange Science* thus dynamically investigate concepts that were and are inherently unstable, and the essays within each section provide a thickly textured analysis of the object of inquiry. As the range of essays within each section shows, the ways in which these scientific topics were addressed by Victorian scientists, artists, and fiction writers alike confound the marginal/mainstream divide in provocative and generative ways. In some cases, although the goal of inquiry was a more scientifically accurate understanding of the natural world, the motivating concerns were primarily spiritual or aesthetic. In others, a scientific idea, whether controversial or established, often served as the catalyst for speculations about the far-reaching implications of scientific discoveries, whether ethical or philosophical. Taken together, these essays demonstrate that, far from existing in a closed system of the pure empiricism, Victorian scientific practice was as affected by imaginative and fantastic possibilities as the fictional works it inspired.

In the first part, "Strange Plants: New Frontiers in the Natural World," the essays explore the complex status of plants in relation to scientific thought and practice in the nineteenth century. Lynn Voskuil opens the collection by examining the unique interspecies discourses that developed surrounding orchids across various scientific, cultural, and colonial contexts. While accounting for orchidology's complicity with British imperialism, she also demonstrates that it offers evidence of a horticultural paradigm that embraced the ontological permeability between the categories of "plant" and "human." She argues that the tendency of Victorian botanists and horticulturalists to perceive orchids as "almost bestial, even human," reveals how Victorian scientists, collectors, and enthusiasts conceived of and construed themselves in relation to the environment. Accordingly, the Victorian fascination with the orchid offers an early instantiation of a trans-species sensibility that would later come to characterize ecological ethics.

Looking at botany in an entirely different context, Meegan Kennedy examines the intellectual and institutional implications of botany's role in a medical education. She uses Edward Forbes's *An Inaugural Lecture on Botany* to illuminate the shifting value of botany as a science at midcentury. Contextualizing Forbes's work within a larger tradition that resisted botany's scientific value due to its cultural associations with sexuality, popular science, and feminine domestic pursuits, she shows how Forbes made a case for botany's practicality as a source of ethical instruction for medical students. Kennedy's essay illuminates the uneven history of scientific virtue by examining an argument for botany's legitimacy that embraced both the morality inherent in an older model of natural theology and an emergent paradigm that privileged pragmatism in medical practice.

Narin Hassan's essay reads the archive of botanical work by Marianne North as an alternate form of scientific practice—one that valued the subjective, sensory experience of the observer over more systematic and institutionalized forms of botanic observation. At a time when botany was increasingly associated with masculine and imperialist agendas, Hassan shows that North's paintings and life writing, which often focus on aesthetic pleasure, offered a significant counterpoint to the dominant discourse of modern science and demonstrated an early sensitivity to environmental ethics in the pursuit of scientific knowledge. Ultimately, Hassan reveals that North resists easy classification across a range of nineteenth-century domains, whether related to gender, travel, science, or empire, thereby challenging the definitional logic underlying these categories.

Taking a sharp turn away from the aesthetic, Elizabeth Chang explores an extraordinary range of late nineteenth-century genre fiction on the subject of killer plants. Reading these fictional accounts of murderous trees, human-absorbing fungus, and alien weeds alongside scientific and philosophic writing about intentionality in plants, she argues that both types of writing were addressing the same complex issue: the possibility of plant consciousness, and how such a phenomenon could be detected or represented. Her essay demonstrates that when writers of all stripes tried to close the gap between the thinking plant and the human, the attempt gave rise to entirely new ontological categories of consciousness that could only be represented at the limits of narrative form.

The second part, "Strange Bodies: Rethinking Physiology," addresses considerations of the human body and its biological functions. Each of these essays examines how bodies could be made to perform in unusual and spectacular ways through science. Danielle Coriale's piece examines Francis Galton's use of psychophysics, the empirical study of the connection between the mind and body by measuring degrees of sensation. Coriale's work brings to light an overlooked aspect of Galton's science, one that recontextualizes and complicates his interest in eugenics. As she shows, in the 1890s, Galton's deafness spurred him to examine how the imagination might supplement stimuli that acted on nerves that otherwise failed to produce sensation; he supported his investigations with examples of auditory imagery taken from nineteenth-century poetry. In addition to examining how scientific thinking can emerge at the intersection of personal and aesthetic experience, Coriale's work offers an early instance of the body being posited as an entity that could be extended past its material limits.

Similarly focusing on bodily communication, James Emmott traces the emergence of what he terms "phonographic physiology," the way in which the human body was, even before the invention of the phonograph, being understood as a biological mechanism that records and replays sound. Emmott begins with an examination of Alexander Melville Bell's development of Visible Speech, a system of writing that records sound through the actions and positions of the mouth. He goes on to trace the conceptual history of the body as a "read-write device" through developments in physiological psychology, vocal pedagogy, and the emergence of "phonographic culture" in the late nineteenth century. His examination culminates in reading George Bernard Shaw's *Pygmalion* as a work that reflects and critiques the disturbing implications of scientific agendas that sought to "civilize" human subjects through physiological reprograming.

Lara Karpenko examines the relationship between mesmeric influence and the infectious popularity of the sensation novel. Focusing on the distinctly physical aspects of mesmerism (through touch) and sensation fiction (through bodily response), Karpenko contends that both science and literature participated in a shared discourse of sympathy that radically posited the possibility of collapsing the boundary between self and other, not despite of, but through, the body. Karpenko goes on to explore how these two strands of sympathy are conflated in Charles Adams's *Notting Hill Mystery*, a work that both replicates and critiques aspects of the sensation novel in its depiction of a series of murders accomplished through mesmeric control. Far from reading it as a sentimental Victorian concept, Karpenko reveals that sympathy was often imagined as a potentially destructive force.

Moving from the body's relationship with the external world to its internal biological functions, Suzanne Raitt contextualizes Oscar Wilde's *The Picture of Dorian Gray* within the scientific discourse on the operations of cell metabolism and division that repairs the waste of the body's natural processes. By foregrounding the novel's considerable engagement with science, Raitt shows how the novel operates as a dark fantasy about the possibility of art substituting for the natural limits of biological processes, and, like those processes, inevitably failing. The picture, then, is not only an aesthetic image of moral repression, but even more universally the literalization of the inexorable biological progress toward death.

The third and final part, "Strange Energies: Reconceptualizing the Physical Universe," moves the volume's focus from an observational study of the natural world to the abstract field of theoretical physics and unseen phenomena. Long before the technological developments of electron microscopes and particle accelerators, the Victorians were tackling the problem of how to detect and demonstrate the structure of unseen forces. Barri Gold identifies one such method as "nonlinear reasoning," a way of thinking about apparently random natural systems. In her reading of "chaotic fictions" by such writers as Alfred Tennyson, Herbert Spencer, James Prescott Joule, and Charles Dickens, she identifies structures that uncannily anticipate key ideas in twentieth-century chaos theory, such as fractals, butterfly effects, and sensitive dependence on initial conditions.

In a similar way, albeit on a subatomic scale, Sumangala Bhattacharya examines the prescient insights of speculative scientific inquiry in her essay on Annie Besant, a controversial freethinker and feminist who, informed by Indian yogic practice, experimentally employed clairvoyant meditation to determine atomic structure. Although the results of Besant's experiments anticipate later discoveries about the nature and structure of subatomic particles, Bhattacharya argues that the more significant value of Besant's work rests in its "critique of the *politics* of scientific authority"—particularly when considering the gendered and colonial implications that underlie such authority. Bhattacharya reads Besant's work as a symptomatic reaction to the institutional barriers to knowledge represented by a masculine, Western scientific establishment. In the context of recent postcolonial and feminist theoretical interventions in science studies, this essay's examination and contextualization of Besant's work explores an early and politically important instance of resistance to the patriarchal, Eurocentric, and anthropocentric discourse of modern science.

Moving from the realm of the atomic to the realm of the literary, Anna Jones's piece examines Edward Bulwer-Lytton's work at the intersection of postmodern theory, Baconian induction, and the occult. In doing so, her chapter challenges the still prevalent tendency in literary studies to discount the value of Bulwer-Lytton's work and his connection to marginal science, despite the fact that both Bulwer-Lytton and these popular sciences had a mass appeal for Victorian audiences. Beyond returning Bulwer-Lytton to his proper intellectual context, Jones even more provocatively shows that his "suggestive system" of intellectual transmission anticipates poststructural literary theory, which similarly takes its point of departure from the decoupling of author and text to determine meaning.

Much has already been said about spiritualism in the Victorian period, but L. Anne Delgado's essay moves away from table-rapping and séances to look at the use of massive data collection by the Society for Psychical Research, a project that aimed to transform the ghost into a fact that might be empirically defined. Her reading of the SPR's *Phantasms of the Living*, a fourteen-hundred-page taxonomic collection of supernatural occurrences, reveals a distinct historical shift in both cultural and scientific registers: culturally, she highlights a turn from popular conceptions of the ghost as a familiar, ancestral presence to one tied to psychological complexity; scientifically, her study emphasizes the Victorian desire to render all aspects of experience, including the otherworldly, statistically knowable. Further, she traces the copious evidence supplied in the compendium to popular literary versions of spiritual phenomena, revealing the ways in which scientific analysis of data transmitted by human interlocutors is always embedded in culture—in this case revealing, perhaps more significantly than any proof of ghosts, a "metaphor for the modern self."

As the volume's subtitle suggests, our essays address Victorian science at the "limits of knowledge," and Tamara Ketabgian's essay approaches this subject at the very outer limit of the knowable universe. She examines the popular and controversial treatise *The Unseen Universe* by the Victorian physicists Balfour Stewart and Peter Gutherie Tait, a work that claimed the second law of thermodynamics provided evidence of another universe being fueled by the entropic waste of our own. Ketabgian argues that this work updated and recast the tropes of natural theology in order to critique science's growing association with secular materialism, thereby mounting an important critique of the way science was conceptualized. She claims that by recasting scientific laws as generalizations, and hypotheses as acts of imagination, Stewart and Tait showed that science as practice is always, in some sense, a heuristic fiction.

Highlighting the epistemological continuum that exists between the nineteenth and twenty-first centuries, Ketabgian's essay functions as a fitting conclusion to our volume. Although *Strange Science* focuses on the nineteenth century, we want to emphasize that such strange and scientific explorations continue today in a number of surprising projects. For example, Nick Bostrom's 2003 "Are You Living in a Computer Simulation?" has given rise to lively and speculative discussions in theoretical physics, while Professor Emeritus William Tiller of Stanford University's Department of Material Sciences has turned his focus entirely to researching psychoenergetics—a field that posits that the human psyche can affect material reality.[28] And, of course, popular debates surrounding global warming and vaccination all hinge on the issue of scientific legitimacy. So while *Strange Science* explores issues and topics that are undoubtedly Victorian, the volume also prompts a reconsideration of twenty-first-century perspectives on scientific thinking and the scientific imagination. At a time when some are questioning the relative value of humanities scholarship and STEM[29] research, *Strange Science* points to the important areas of intersection between scientific, humanistic, and artistic endeavors, intimating fresh ways of aligning the categories of knowledge that organize the academy today. Moreover, and perhaps even more poignantly, our collection reveals the permeability between the mundane and the extraordinary, suggesting that the "strange" always tantalizingly remakes our everyday reality.

Notes

1. Henry Dircks, "Science versus Spiritualism," *The Times*, December 27, 1872.

2. Ibid., emphasis added.

3. Ibid. It is worth noting that Dircks does not quite acknowledge that one-tenth of spiritualist séances may be scientific; rather, he asserts "that nine-tenths of the séances" are antiscientific. Still, he does seem to allow for some portion of séances to qualify as scientific practice.

4. The "ghost" is named after fellow scientist and coinventor John Henry Pepper. James Secord, "Quick and Magical Shaper of Science," *Science* 297 (2002): 1648.

5. Mary Poovey, *A History of the Modern Fact: Problems of Knowledge in the Sciences of Wealth and Society* (Chicago: University of Chicago Press, 1998). Page numbers for subsequent citations will be given parenthetically in the text.

6. Evelyn Fox Keller and Helen E. Longino, introduction to *Feminism and Science*, ed. Evelyn Fox Keller and Helen E. Longino (Oxford: Oxford University Press, 1996), 1.

7. Mel Y. Chen, *Animacies: Biopolitics, Racial Othering, and Queer Affect* (Durham, NC: Duke University Press, 2012), 4.

8. Ibid., 3.

9. "Paraffine," *All The Year Round* 1 (1868): 58; Charles Neidhard, "Review: On Certain Medical Delusion," *Homeopathic Examiner* 3, no. 2 (1842): 352.

10. Bruno Latour, *Pandora's Hope: Essays on the Reality of Science Studies* (Cambridge: Harvard University Press, 1999), 15.

11. See, for instance, Steven Shapin, "Phrenological Knowledge and the Social Structure of Early Nineteenth-Century Edinburgh," *Annals of Science* 32, no. 2 (1975): 219–43; Roy Wallis, ed., *On the Margins of Science: The Social Construction of Rejected Knowledge*, Sociological Review Monograph 27 (Keele: University of Keele, 1979); and Roger Cooter, *The Cultural Meaning of Popular Science: Phrenology and the Organization of*

Consent in Nineteenth-Century Britain (Cambridge: Cambridge University Press, 2005).

12. Thomas Gieryn, "Boundary-Work and the Demarcation of Science from Non-science: Strains and Interests in Professional Ideologies of Scientists," *American Sociological Review* 48, no. 6 (1983): 782.

13. Some representative titles: Wallis, *Margins of Science*; W. F. Bynum and R. S. Porter's *Medical Fringe and Medical Orthodoxy, 1750–1850* (London: Croom Helm, 1987); Martin Fichman, "Biology and Politics: Defining the Boundaries," in *Victorian Science in Context*, ed. Bernard Lightman (Chicago: University of Chicago Press, 1997), 94–118; Evelleen Richards, "Redrawing the Boundaries: Darwinian Science and Victorian Women Intellectuals," in Lightman, *Victorian Science in Context*, 119–42; Roy Wallis and Peter Morley, eds., *Marginal Medicine* (New York: Free Press, 1976); Thomas Gieryn, *Cultural Boundaries of Science: Credibility on the Line* (Chicago: University of Chicago Press, 1999); Alex Warwick, "Margins and Centres," in *Repositioning Victorian Sciences: Shifting Centers in 19th Century Thinking*, ed. David Clifford, Elisabeth Wadge, Alex Warwick, and Martin Willis (London: Anthem Press, 2006), 1–16.

14. Roger Luckhurst, *The Invention of Telepathy* (Oxford: Oxford University Press, 2002); Alison Winter's *Mesmerized: Powers of Mind in Victorian Britain* (Chicago: University of Chicago Press, 1998).

15. While *Pseudo-science and Society in Nineteenth-Century America*, ed. Arthur Wrobel (Lexington: University Press of Kentucky, 1987) addresses a range of marginal sciences without focusing on the construction of legitimacy, it does so only in the U.S. context, concentrating on how these sciences supported the emergence of a uniquely American national identity.

16. Alison Winter, "Construction of Orthodoxies and Heterodoxies in the Early Victorian Life Science" in Lightman, *Victorian Science in Context*, 31.

17. Jack Morrell and Arnold Thackeray, *Gentlemen of Science: Early Years of the British Association for the Advancement of Science* (Oxford: Clarendon Press, 1981), 20.

18. Ibid., 276–80, 287–91, 515.

19. As Susan Cannon has famously argued, in the early Victorian period, science and culture cooperated in developing and supporting a singular "Truth Complex," a scientific worldview largely endorsed by theologians, philosophers, and literary figures alike. *Science in Culture: The Early Victorian Period* (New York: Science History Publications, 1978).

20. Jay Clayton, *Charles Dickens in Cyberspace: The Afterlife of the Nineteenth Century in Postmodern Culture* (Oxford: Oxford University Press), 82–83.

21. Laura Otis, introduction to *Literature and Science in the Nineteenth Century: An Anthology*, ed. Laura Otis (Oxford: Oxford University Press), xix.

22. Cary Wolfe, *What Is Posthumanism?* (Minneapolis: University of Minnesota Press, 2010), xii.

23. Deborah Denenholz Morse and Martin Danahay, introduction to *Victorian Animal Dreams: Representation of Animals in Victorian Literature and Culture*, ed. Deborah Denenholz Morse and Martin Danahay (Aldershot: Ashgate, 2007), 3–4.

24. Forianne Koechlin, "The Dignity of Plants," *Plant Signaling and Behavior* 4, no. 1 (2009): 78.

25. Jim Endersby, "Sympathetic Science: Charles Darwin, Joseph Hooker, and the Passions of Victorian Naturalists," *Victorian Studies* 51, no. 2 (2009): 299–320; Winter, *Mesmerized*.

26. Keller and Longino, *Feminism and Science*, 1.

27. Mary Louise Pratt, *Imperial Eyes: Travel Writing and Transculturation*, 2nd ed. (New York: Routledge, 2008).

28. Nick Bostrom, "Are You Living in a Computer Simulation?" *Philosophical Quarterly* 53, no. 211 (2003): 243–55. For more on the scientific response to Bostrom's work, see Silas R. Beane, Zohreh Davoudi, and Martin J. Savage, "Constraints on the Universe as a Numerical Simulation," *European Physical Journal A* 50, no. 9 (2014): 1–9. For more on Tiller's work on psychoenergetics, see William A. Tiller and Walter E. Dibble, "A Brief Introduction to Intention-Host Device Research," 2009, William A. Tiller Institute for Psychoenergetic Science, accessed April 12, 2015, http://tiller-institute.com/pdf/White%20Paper%20I.pdf. In this article, Tiller claims that he and his colleagues have been able to alter the acid/alkaline balance of water "by creating an intention to do so" (2).

29. STEM (science, technology, engineering, and mathematics).

PART I

Strange Plants

New Frontiers in the Natural World

CHAPTER 1

Victorian Orchids and the Forms of Ecological Society

Lynn Voskuil

In "The Strange Orchid," one of his late-century stories, H. G. Wells portrays the relationship between an orchid fancier and his newly acquired specimen. With dystopic finesse, Wells imagines a plant that expresses malicious intent and aggressive agency, one whose "tentacle-like aërial rootlets" eventually grow strong and long enough to choke the horticulturist in its parasitic grasp. While the fancier escapes with his life and the orchid expires "black . . . and putrescent," the story plays on the popular Victorian perception of orchids as almost bestial, even human, in their forms and habits of growth—and thus (like humans) capable of inexplicably strange behavior.[1] This perception was shaped by a century-long tradition of "orchidology"—an enormous body of work accumulated by botanists, plant hunters, commercial growers, and ordinary gardeners—that explores, often with great sophistication, the morphology, physiology, and ecology of orchids. Throughout the nineteenth century, orchids appeared with increasing frequency in daily British life, with their own "orchid houses" and a growing population of enthusiasts dedicated to their care and feeding. Fascinated more by orchids than by any other plant, Victorians were attracted in great part to their apparent sensitivity, their capacity for responding dramatically to other elements in their environments, including humans. The minutely scripted, even intimate, ecological relationships between horticulturists and their orchids capture the readiness of many Victorians to conceive of boundaries between species as fluid rather than absolutely fixed. By 1898, when Wells's story was published, orchids per se would thus no longer have been considered "strange." For fin de siècle readers, the horror of this story may well have resided more in its representation of botanic malevolence than in its violation of the limits between human and nonhuman species.

The fascination with species boundaries evident in Victorian orchid literature, including Wells's story, urges a precise grasp of the disciplinary frameworks put into play when we analyze these texts because the paradigms most applicable in this case are themselves variably concerned with cultural, epistemological, and ontological boundaries. On the one hand, the nineteenth-century, large-scale importation of orchids may be analyzed as an aspect of the British imperial project, with orchids as commodities that elevated Britain economically and as botanic objects that were perceived to confirm its cultural and scientific fitness for global rule. This paradigm, with its roots in the work of Edward Said, is premised on an awareness of the inflexible boundaries Britain erected between itself and its colonial "others," whether those "others" are understood to be indigenous peoples, conscripted animals, or plants ripe for plunder.[2] On the other hand, the readiness among Victorian horticulturists to rupture ontological boundaries may be interpreted as a prescient example of interspecies awareness, as evidence that Victorians were more intellectually and morally complex than has been historically acknowledged. This paradigm, grounded in the recent innovations of posthumanist studies, views boundaries between species as permeable, even continuous, and celebrates that fluidity as ethically commendable.

Both frameworks are invoked in this essay, but neither is implemented unskeptically because the complexities of human-orchid relationships in Victorian Britain elude the explanatory structures of these frameworks as they have been configured today. Rather than wholly conforming to either, Victorian orchid literature redirects our focus to the contours of ecological relationships as those were understood in the nineteenth century. At the same time that Victorians imagined themselves as bonded to orchids with affective singularity, they also envisioned particular roles for themselves, human horticulturists, in their understanding of what was then the emerging science of ecology. Victorian orchid literature thus organizes conceptual boundaries—between people and plants, between Britain and its colonies, between nature and culture—with an ideological flexibility that is unexpected and transcends its historical moment.

Such flexibility is consistent with new debates in the humanities that seek not merely to shift or blur boundaries between species but to question them altogether—along with the assumption of human exceptionalism that mandates such boundaries in the first place. Jane Bennett in particular has argued, with force and efficiency, for a notion of agency that no longer privileges human intention and will. "A lot happens to the concept of agency," as she puts it, "once nonhuman things are figured less as social constructions and more as actors, and once humans themselves are assessed not as autonoms but as vital materialities."[3] While Victorian orchid literature underscores the mutually constitutive effects of empire and environmentalism, then, it also shows how Victorian conceptions of other species might address the inadequacies of some current heuristic paradigms, most notably the boundaries that position plants and people in ranked relations to each other. Strange as it may seem, Victorians and their orchids may well have much to say about our own ecological and disciplinary commitments today.

Orchid Ontology

The nineteenth century may be thought of as the century of the orchid, at least in the West. Lewis Castle, a Victorian historian of orchids, provides some rudimentary figures on its early introduction and cultivation in Britain. The first exotic orchid arrived in Britain in 1731, he says, as a dried specimen that was resuscitated; by the middle of the eighteenth century, there were still just four nonnative orchid species cultivated in Britain. Knowledge of exotic orchids was thus very limited, writes Castle, until Linnaeus published the second edition of *Species Plantarum* in 1763, in which he enumerated ninety-one species (itself far short of the hundreds of genera and thousands of species now classified as members of this family).[4] But by the end of the eighteenth century, notes Castle, there were approximately fifty species in British cultivation of both exotic and native origin.[5] These numbers grew quickly in the early nineteenth century. Whatever the accuracy of Castle's figures, they capture the remarkable intensification of interest in exotic plants that was fueled by growing numbers of introductions into Britain in the late eighteenth and nineteenth centuries. The Horticultural Society of London (later the Royal Horticultural Society) was founded in 1804 and soon began sponsoring plant-hunting expeditions to secure new, exotic specimens for British gardens, while the Royal Botanic Gardens at Kew also imported numerous new species for scientific study. Where orchids in particular were concerned, early nineteenth-century collections, established initially by aristocratic fanciers with the means to fund their pursuits, were instrumental in solidifying orchid culture in Britain. Large, commercial nurseries soon began funding their own plant hunters and importing their own orchid stock; Conrad Loddiges and Sons opened a nursery in Hackney in the early part of the century, followed by James Veitch and Sons with the Royal Exotic Nursery in Kensington and Benjamin Samuel William with the Victoria and Paradise Nursery in Holloway—all of which made orchid culture possible for the average, middle-class, and (eventually) even working-class gardener. By 1840, the enthusiasm was already intense, leading collector James Bateman to proclaim that an "*Orchido-Mania* . . . now pervades all classes . . . to a marvelous extent."[6]

The nineteenth-century orchid literature that documents this "mania" is large and variable. Orchids were cataloged and described in both horticultural and botanical sources; this mix reflects the blended intellectual culture of nineteenth-century Britain, when the practices of horticulturists and botanists still overlapped to a great degree and science was often a popular pursuit. The career of John Lindley is exemplary in this regard, with its orientations toward both amateur gardeners and professional botanists. He was instrumental in classifying and describing newly introduced orchid species in the 1830s and 1840s, with volumes like *Folio Orchidacea* analyzing them for more scientifically inclined readers, and others, like *Sertum Orchidaceum*, targeting readers interested in the aesthetic qualities of orchids.[7] Journals that featured exotic plants—*Curtis's Botanical Magazine*, for example, and the *Botanical Cabinet*—spread the word about many newly introduced orchid species, and later in the century, periodicals like the *Orchid Review* and *Orchid Album* focused exclusively on orchids; general gardening magazines like the *Gardeners' Chronicle* (cofounded by Lindley) frequently ran articles on orchid cultivation.[8] Finally, many plant hunters published memoirs about their adventures that featured harrowing narratives, while scores of growers published manuals of orchid care aimed at the general reader.

Like the thousands of other exotic plants imported into eighteenth- and nineteenth-century Britain, orchids may be understood as artifacts of empire. Scholars have amply demonstrated that plants figured in the Western imperial mission, a mission that included the uses of botanic language and taxonomic systems to promote "European global expansion and colonization."[9] Predictably, many orchid sources betray an imperialist sensibility, revealing their contributions to the popular diffusion of

Fig. 1.1. *Dendrobium Albo-Sanguineum, Curtis's Botanical Magazine* 85 (1859): Tab. 5130. (Courtesy Huntington Library.)

empire characteristic of nineteenth-century culture at large. Plant hunters, for instance, often exhibited a blatant disregard for the effects of large-scale plunder in their pursuit of lucrative species. One collector, in search of *Odontoglossum* orchids in a dense Andean forest, describes the methods he used to secure specimens "high up out of reach of the native climbers." With his goal of gathering as many plants as possible, he "provided [his] natives with axes and started them out on the work of cutting down all trees containing valuable orchids." After about two months' work, he concludes, "we had secured about ten thousand plants, cutting down to obtain these some four thousand trees, moving our camps as the plants became exhausted in the vicinity."[10] Such a sensibility was, of course, one of the primary engines of empire in the nineteenth century, and orchid fanciers were not immune to it. And with its emphasis on British exceptionalism—the conviction that Britain in particular was uniquely qualified, even obliged, to discover and plunder the natural resources of other global regions—this sensibility relies on the erection of firm cultural boundaries and hierarchies between the British Empire and the people and resources it colonized.

At the same time, orchid literature bespeaks competing attitudes that challenge such boundaries, attitudes that manifest an early ecological awareness of human engagement with other species and a different sense of the boundaries between them. This emerging awareness laid the groundwork for conceptions of a social life organized around ecological alliances rather than, say, kinship ties or social contracts; and the concept of ecology itself opened the door to the idea of social relationships between humans and nonhuman organisms and things. The term "oecologie" was coined by German zoologist Ernst Haeckel in 1866 to name and advance a new science "of the relations of the organism to the environment including, in the broad sense, all the 'conditions of existence.'" In Haeckel's formulation, these "conditions" could be either organic (other organisms) or inorganic (climate, nutrients, surrounding physical and chemical elements).[11] Either way, his understanding of "ecology" was strongly Darwinian in its adaptations of the "economy of nature"—a term Darwin himself derived from Linnaeus—among other concepts.[12] While these ideas are central to *On the Origin of Species,* they also figure significantly in Darwin's *The Various Contrivances by which British and Foreign Orchids Are Fertilised by Insects,* a volume that he considered to be an evidentiary foundation for certain points in *Origin* but that was also taken up by many orchid enthusiasts and referenced in many orchid manuals.[13] By 1860, then, well before Haeckel coined his term, the general sense of "ecology" was already in wide circulation in Britain, as the study of how organisms interact with each other and additional elements in their environments, including humans. And the mainstream popularity of this idea—its currency with gardeners and farmers as well as botanists and zoologists—guaranteed a degree of practical awareness and cultural diffusion that would have eluded a more strictly scientific dissemination of the concept.

One of the primary ecological markers for orchid enthusiasts was the effects of orchids on themselves, effects that promoted the awareness of an interspecies exchange with crucial impact on humans. Perhaps no other botanic family was perceived to touch its growers so palpably—in both physiological and affective terms—as the orchid. This effect is evident, first of all, in the episodes of discovery in orchid-hunting narratives, accounts of the moment when years of pursuit and travail are finally rewarded by the location of a rare specimen. As something of a set piece in these texts, these accounts often stress the moment of discovery as a sensory-laden experience that transports and sometimes even overwhelms the seeker. In one late-century narrative, for example, an orchid hunter is simultaneously seduced by the brilliant coloration of massed blooms and overcome by their putrid smell in his pursuit of an ultimately unattainable specimen. Lured by accounts of a "demon flower" deep in the Amazon rain forest, he pushes his team forward for weeks, only to have three of them eventually fall senseless in response to a "peculiar sickening odour pervading the heavy, heated air." The odor is the scent of the "great mass of Orchids," a glamorously colored species that was bending the trees and plants with its heavy, refulgent weight.[14] The "demon flower" finally could not be collected, its exhalations preventing anyone from approaching it closely. Other accounts note mammoth orchids with blooms far larger than any cultivated in Britain; caches of plants where they were least expected to be found; and "immense clumps" that astonished "even the most stoical observer."[15]

These episodes often serve as narrative climaxes in orchid-hunting texts where suspense is structured around botanic discovery, a function that led to heightened sensory language. But they also underscore the disorienting effects of orchids on humans, drawing on a convention of naturalist writing about the tropics that had been in use at least since Alexander von Humboldt published his *Personal Narratives of Travels to the Ecquinoctial Regions of the New Continent* in the early nineteenth century. In this text, Humboldt expresses a destabilizing sense of scale and quantity when he encounters the lushness of the tropical forest.[16] The episodes of discovery in later orchid literature reprise these scenes in provocative ways, emphasizing not only the stupefaction of travelers unaccustomed to rain forest habitats but also the vigorous, even forcible, habits of orchid growth. The "demon flower" exemplifies these traits with particular clarity, its fetid smell forbidding approach and preventing its transport to Britain. In similarly compelling ways, other specimens seemed to lure collectors to them with their "uncanny" features.[17]

Recalling the qualities of Wells's "strange orchid," these traits were bound up in the variable and fantastic shapes of orchid form, form that was experienced as assertive and even communicative by Victorian growers. The term "form" appears with striking frequency in orchid sources. Orchid fanciers were astonished by the "endless varieties of form" that orchids assumed, and their professions of astonishment became a convention of orchid literature.[18] Darwin himself made a reflexive nod to this practice on the very first page of his own volume when he noted, "Orchids are universally acknowledged to rank amongst the most singular and most modified forms in the vegetable kingdom."[19] To some degree, the attention to form reflects the reach of the science of "morphology," which Darwin called "the most interesting department of natural history," and it is no surprise that the term "form" also appears frequently in *Origin*.[20] The mention of form in the orchid volumes, though, is more than a convention. More crucially, the obsessive focus on the intricate and variable forms of orchids in horticultural and botanic literature shows how this botanic family figured in nineteenth-century ecological thought. In the formal variety of orchids, botanists and growers found not only scientific and aesthetic curiosities but also the evidence for different forms of response to the conditions of existence that orchids experienced. The variable forms of orchids were perceived in turn to affect their growers in different ways, with fancier and bloom both shaped by the mutually constitutive ecological relationship.

The earliest nineteenth-century collectors were immediately receptive to what Bateman called the "magic influence" of orchid form. For him, orchids represented an "encroachment" on the animal kingdom, so potent was their mimic capacity.[21] For Frederick Boyle, their readiness to mutate resulted in "glorious freaks" that were seen in no other "realm of [nature's] domain."[22] Even Lindley, ambitious to establish botany as a professional science, lapsed into colorful prose when describing "the extremely remarkable forms of some species."[23] Of the microscopic *Oberonia rufilabris,* he wrote (echoing Bateman), they are "all so different from other plants that one might almost doubt their even belonging to the vegetable world. . . . Pythagoras would have found living evidence of animals transmuted into plants."[24] The structure of *Cynoches maculatum* amazed him even more. "Did any one ever see such a flower before?" he wondered. "Which is the top, which is the bottom? What are we to call that long club foot, which is cloven too; and what the crooked fingers daggled with blood, which spread from the middle of one of the leaves, as if about to clutch at something? And what moreover *can* they all be *for*?"[25] The qualities attributed to orchid form—assertion, compulsion, mimicry, sensuality, even (for Wells) agency—led some orchid fanciers

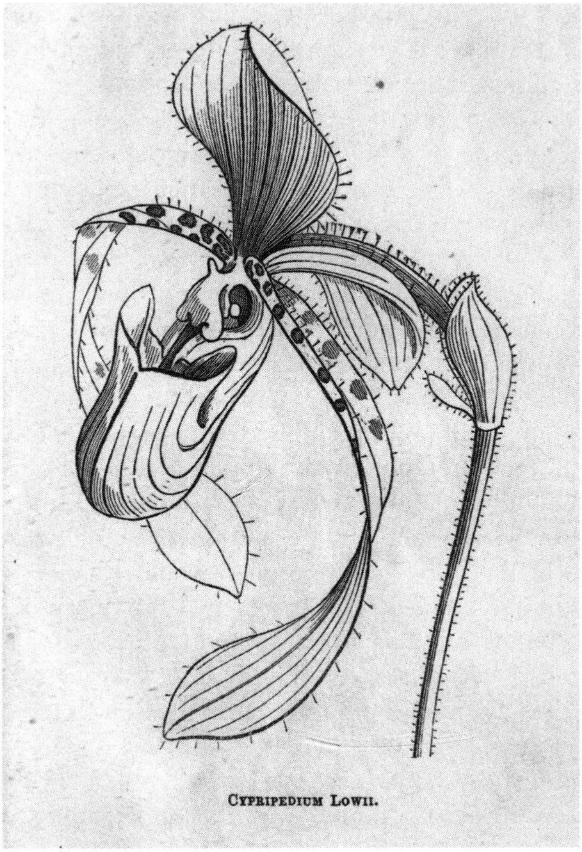

Fig. 1.2. *Cypripedium Lowii* with face-like structure. Thomas Appleby, *The Orchid Manual, for the Cultivation of Stove, Greenhouse, and Hardy Orchids* (London, [1865]): n.p. (Courtesy Huntington Library.)

to take the next step and imagine their plants not merely as bestial but as hominid.[26] "Their intelligence is almost human," wrote one enthusiast.[27] Another interpreted their form as akin to the human face. "The element, the base, the constituent idea of an orchid is that of a life, of organization, of a being, of a face with all its parts, its line and expression," he wrote. "They look at us, indeed they have faces, and so many thousands and hundred thousands of orchids with which I have been face to face, I never yet tired to again and again study the character of their kind. They have faces."[28]

This horticultural language anticipates several strands of posthumanist studies today, most notably the blurring of species boundaries and the emphasis on alternative epistemological models. As Cary Wolfe puts it, posthumanism "fundamentally unsettles and reconfigures the question of the knowing subject and the disciplinary paradigms and procedures that take for granted its form and reproduce it."[29] For Wolfe, this shift marks a recent and profound break with previous, deeply entrenched ways of understanding epistemology and subjectivity. Paul Outka, however, has located the roots of posthumanist studies not in contemporary animal studies (as Wolfe does) but in the nineteenth-century embrace of materialism, specifically in "episodes when an individual experienced and recorded an often profoundly disconcerting awareness of the radical material identity between his or her embodied self and the natural world." For Outka, these episodes demonstrate that a "nascent posthuman consciousness" emerged far earlier than contemporary posthumanists allow, thereby exposing the "ahistorical hubris" and radical claims of newness characteristic of some versions of current posthumanist thought.[30] Orchid-human relationships in the nineteenth century manifest several aspects of this emerging sense of oneness with material nature. While those relationships may thus be understood as an example of the now-familiar Enlightenment practices of recording and documenting the natural world, including its difference from and submission to the human, they were also encounters that troubled the certainties that separated "civilized" or "evolved" humans from other living things. This latter quality may be traced in particular in the impression of intimacy these relationships registered, the sense that human and orchid development is inextricably intertwined.

This sense was grounded in what was, by the 1860s, respectable science, most notably the concept of coevolution as Darwin explained it particularly in *The Various Contrivances by which Orchids are Fertilised by Insects*.[31] The volume provides episode after intricate episode of insects and orchids evolving together to ensure the survival of both animal and plant. Darwin's description of *Angraecum sesquipedale*, an orchid native to Madagascar, famously illustrates this principle. Puzzled by the existence of a nectary in this orchid of "disproportionate length," he eventually hypothesized a large moth with a long proboscis as the agent of fertilization, pointing out that the extinction of either orchid or moth would entail the extinction of the other.[32] In this case, the formal characteristics of both species had evolved in precise, even intimate relation to the other. Like Lindley, his fellow botanist, Darwin found orchid form astonishing in its variability and especially in its capacity to compel ecological interaction. Following a long, intricate discussion of the pollination process of *Orchis pyramidalis*, for example, he describes the attractions of the bloom for the moth that enables fertilization. "As the flowers are visited both by day and night-flying Lepidoptera," he says, "it is not fanciful to believe that the bright-purple tint (whether or not specially developed for this purpose) attracts the day-fliers, and the strong foxy odour the night-fliers." He goes on to describe how the long nectary of *O. pyramidalis* requires the visiting moth to suck nectar slowly and thus take on a large, firmly attached load of pollen before it leaves one bloom to visit another. The orchid's form and properties thus enables the insect to "effect a union between two distinct individuals."[33] In the case of *O. pyramidalis,* visiting Lepidoptera are lured by color, odor, and form, much as the orchid hunter had been seduced by the "demon flower" in the Amazon rain forest; the narrative of this orchid's pollination, in other words, foregrounds the experience of the moth in much the same way that it was emphasized for the orchid hunter in the adventure tale.

For many Victorian fanciers, it was but a short step from orchid-insect relationships to orchid-human relationships, especially after the process of orchid hybridization was discovered in the 1850s. John Dominy, a gardener at the Veitch nurseries, began experimenting with the process in the early 1850s and brought the first hybrid to flower in 1856.[34] This event is significant because, in the act of hybridizing, the human grower manually places the pollen from one plant on the stigma of another, thus replacing the insect agent in the process of pollination and becoming even more intimately involved in the orchid's life cycle. The process of hybridization alerted cultivators to habits of orchid form and growth that seemed to confirm their almost human quality. Boyle, for example, tells the story of a fancier who "amused himself with investigating the structure of a few Cypripeds, after reading Darwin's book, and he impregnated them. To his astonishment, the seed-vessel began to swell," with the grower assuming he would soon have viable seeds. Unfortunately, Boyle continues, he did not yet know that "pseudo-fertilization can be produced, actually, by anything. So intensely susceptible is the stigmatic surface of the Cypriped that a touch excites it furiously . . . it will go sometimes through all the visible process of fecundation . . . but, of course, there is no seed."[35]

Clearly, this eroticized strain of horticultural writing genders and sexualizes the relationship between specimen and fancier. Related examples of eroticized style are also evident in Darwin's work, including an account of the *Catasetum* orchid, a species that ejects its pollinium so forcibly as to shoot it "to the distance sometimes of two or three feet."[36] Often, such discourses confirm traditional ideologies of sex and gender, as Boyle's text demonstrates: the (male) grower inserts pollen into the (female) bloom, exciting the "susceptible" stigma so "intensely" that the flower behaves as if "impregnated."[37] At the same time, much orchid literature plays into the tendencies of nineteenth-century writing in general to exoticize and orientalize non-Western regions and peoples. This tendency is conspicuously illustrated in some of the orchid texts already cited here, texts that emphasize the exotic provenance of some orchid species, their sensual appeal, and the habits of growth that distinguish them from more sedate Western plants. These features of orchid literature reinforce scholarship of the past several decades, work that has persuasively illuminated the entangled ideologies of race, gender, and empire in nineteenth-century Western texts.

Here again, however, orchid literature is twofold and contradictory: it confirms our twentieth-century scholarship about British representation of race, gender, and empire on many fronts, yet it also makes available other circulating explanatory paradigms, most notably those that question the subordination of plants to people and the maintenance of stringent boundaries between them. Victorian orchid growers were so fascinated by orchid sex because it violated their expectations of both botanic behavior and botanic ontology. While there were biological explanations for the false pregnancy that Boyle observed in his excitable orchid, it seemed to confirm the sense of volition, even duplicity, in plants, and some Victorian scientists began to theorize provocatively along these lines. W. Lauder Lindsay, for example, argued that "some form or degree of Consciousness exists in plants" and built on Darwin's work in *The Movements and Habits of Climbing Plants* to analyze certain botanic behaviors as examples of "choice," "selection," and "preference."[38] He also speaks to the kinds of ecological interactions that horticulturists observed practically. "*Attachment* to place or things, which are obvious in the case of many climbers, may, perhaps, in other plants explain much that the botanist, horticulturist, aboriculturist, floriculturist, or agriculturist cannot otherwise satisfactorily account for," he observed. "Plants exhibit occasionally *individuality*, and even *eccentricity*, for which we cannot account, any more than we can for similar peculiarities in man or other animals."[39] For orchid fanciers closely attuned to the preferences of their own plants, such concepts seemed only to tighten the affective tie that they believed bound their orchids to them. "It is said of Orchids that they, like domestic animals, soon find out whether they are under the care of one who is really fond of them, and that they respond by thriving or failing according," wrote W. H. White. He also argued that this care must be "unfeigned" by the grower—because, presumably, the orchid could see through counterfeit affection.[40]

What is at stake in such seemingly eccentric pronouncements from little-read nineteenth-century horticultural literature? They challenge, first of all, the theoretical confines of species, a challenge that Victorians themselves recognized. Once the process of hybridization was perfected, it was found to be a straightforward practice, reinforcing the conclusions of those horticulturists who questioned the species boundaries that botanists had erected taxonomically. Following Dominy's initial success, Donald Beaton, an eminent gardener and horticultural journalist, reportedly observed, "There is not such a thing in nature as a species as meant by botanists."[41] If the creation of new hybrid species and even genera was so simple, how should these categories be conceived? The apparent cooperation of orchids in the process—and the intimate effects of specimen and grower on each other—reinforces such questions and raises new ones about agency and subjectivity. Mel Chen has recently explored related issues in her discussion of "animacies," drawing on Bennett's conception of "vibrant matter" to question long-standing Cartesian divisions between mind and body. "It is possible," Chen insists, "to conceive of something like the 'affect' of a vegetable, wherein both the vegetable's receptivity to other affects and its ability to affect outside of itself, as well as its own animating principle, its capacity to animate itself, become viable considerations."[42] Victorian orchids and their growers, as both are described in nineteenth-century horticultural literature, exemplify such affective mutuality.

Orchid Ecology

Even as Victorian orchid literature raises significant questions about species relationships and ontology, it does not wholly relinquish the dominant, or at least differentiated, role of humans in ecological exchange with plants. Lindsay's reflections about "mind" in plants, for example, clearly anthropomorphize botanic behaviors, while the notion of facial form in orchids explicitly patterns their structure on human form. Bennett has argued that the penchant for anthropomorphizing nonhuman organisms and things is not necessarily a mark of human-centered thinking—and in fact can promote the practice of thinking across ontological boundaries. "A chord is struck between person and thing," as she puts it, with the human "no longer above or outside a nonhuman 'environment.'"[43] Such challenges to humanist ideas may be functional in nineteenth-century orchid literature, at least to a certain degree.

But the imperialist framework of nineteenth-century global operations made it very difficult to question the certainty of British—and human—exceptionalism, with the result that horticulturists retained a firm, dominant role in most orchid literature. The tendency to anthropomorphize orchids thus largely preserved the centrality of humans in the ecological imaginary, understanding the orchid (and, in some cases, other plants as well) as if it were, at most, a subordinate hominid form. Much orchid literature thereby extended human ways of being-in-the-world.

For at least some Victorian horticulturists, however, human exceptionalism entailed human responsibility as well. These horticulturists began to ponder the particular responsibilities of humans in a rapidly changing global landscape, a concern that mitigated the imperialist impulse to some degree. Such viewpoints are less visible to traditional historians of botany like Drayton, who tend to focus on the rise of professional science and state-sponsored botany, devoting little attention to amateur horticulture. It is the horticultural literature, however, that manifests these attitudes most obviously because gardeners and nurserymen were closely attuned to the conditions under which individual plants grew and thrived; they were thus particularly sensitive to the effects of their own actions on the plants they collected and cultivated. Orchids underscored these ecological concerns because they responded so dramatically to human intervention in both their artificial and indigenous habitats. In the process of working through some of these issues—in both books and gardens—orchid fanciers in particular developed sophisticated forms of ecological awareness that suggest new models of agency and human responsibility.

By the final two decades of the nineteenth century, a number of writers were already lamenting the despoliation of orchid habitats in the wild. Frederick Boyle, a well-known fancier who published several books on orchids, wrote extensively on this issue. "The English demand has stripped whole provinces," he notes, "and now all the civilized world is entering into competition." While some species of orchids repropagate freely, others—like *Odontoglossum,* one of the most prized Victorian genera—grow slowly and are difficult to cultivate from seed. Boyle was thus very concerned "that Odontoglossums [*sic*] carried off will not be replaced for centuries."[44] Boyle also commented on the practice of razing mature exotic hardwoods to secure orchids in their uppermost branches. "It is a terribly wasteful process," he observes. "If we estimate that a good tree has been felled for every three scraps of Odontoglossum which are now established in Europe, that will be no exaggeration. And for many years past they have been arriving by hundreds of thousands annually!"[45] Boyle even had the prescience to consider the situation in decades to come in his discussion of *Cypripedium,* a genus that is far easier to cultivate and hybridize than *Odontoglossum.* Noting the disappearance of several species of this orchid in the wild, he emphasized the loss as a "serious warning." "In seventy years we have destroyed the native stock of two orchids, both so very free in propagating that they have an exceptional advantage in the struggle for existence," he lamented. "How long can rare species survive, when the demand strengthens and widens year by year, while the means of communication and transport become easier over all the world?"[46] Here ecological sensitivity is paradoxically driven by human acquisitiveness: to love and cultivate an orchid necessitated an awareness of the larger ecosystems in which they flourish.[47]

This burgeoning ecological awareness motivated both the preservation of indigenous systems and the creation of artificial ones in new global regions, involving entities as large as the British government and as small as the individual fancier. For Boyle, one important response to the threat of extinction was government intervention, a step he called for on more than one occasion.[48] But he and other growers also explored the ecological significance of artificial cultivation on the future of orchids as a global botanic family. Virtually every British orchid manual—whether or not it professed environmental awareness and concern—offered advice to the home grower about the conditions that would enable exotic orchids to thrive in the British climate.[49] Some growers were even consciously and acutely aware of the role horticulturists and hybridists might be called upon to play in a future world where orchid habitats in the wild had been destroyed. Echoing Boyle, William Watson observed, "Though the process of extinction may be slow, it is sure. It is thus, then, that we shall have to depend on the work of the cultivator to retain the species by raising them from seed, as well as by procuring new sorts by means of cross-fertilization."[50] He followed these observations with detailed instructions on hybridizing orchids, sowing seeds, and potting up the seedlings. Boyle himself mitigated his own dismal predictions of orchid extinction by imagining a world where orchids evolved to rely on human intervention. His case study focused on the genus *Cypripedium,* the easily cultivated orchid whose near extinction in the wild he had previously lamented. Drawing on Darwin, he noted that this genus is a primitive one, incapable of self-fertilization and attractive to few insects that can easily fertilize it. "Its time has passed—Nature is improving it off the face of the earth," Boyle observed. In response, he accorded human cultivators a special role in its preservation. "A gradual change of circumstances makes it more and more difficult for this primitive form of orchid to exist, and, conscious of the fate impending, it gratefully accepts our help."[51] In this view, horticulturists have the opportunity not merely to right a wrong—to preserve a species whose habitat humans had destroyed—but even to involve the orchid in new modes of coevolution with humans. "Darwin taught us to expect," Boyle concluded, "that species which can rarely hope to secure a chance of reproduction will learn to make the process as easy and as sure as the conditions would admit—that none of those scarce opportunities may be lost. And so it proves."[52] The ease with which the genus *Cypripedium* could be hybridized and germinated demonstrated to Boyle that at least one orchid had already coevolved with humans to the point where its seemingly certain path to extinction had been reversed.

Such ideas are open to variable interpretations. One, of course, is that orchid collecting and cultivation served only to confirm the Victorian impulse to colonize the world—including the botanic world that had expanded so rapidly and exponentially in the nineteenth century. The practical knowledge developed by horticulturists, however, accentuates the complex affiliations of empire and environmentalism and shows that the collection of exotic species promoted ecological awareness even as it satisfied certain territorial appetites. The view of orchids as individual specimens with affective sensibilities, for example, prompted horticulturists to imagine themselves as sympathetically and ecologically linked to their plants in a mutually constitutive relationship. However strange this may seem, it anticipates recent botanical research that analyzes "plant intelligence" and conceives of plants not as passive beings but as "behavioral organisms with a capacity to receive, store, share, process, and use information from the abiotic and biotic environments."[53] This research, like the Victorian research that preceded it, has reorganized our awareness of how humans and animals interact with plants and share global space with them, thereby mounting a challenge to what Robert Markley has recently described as "a kind of eukaryotic provincialism that reinscribes a host of self-congratulatory assumptions and values about *homo sapiens* as the shepherd, manager, and conservator of the planet's biota."[54] While Victorian growers may still have placed themselves at the center of the horticultural universe, their vast knowledge of orchids—including the many environmental elements and practices that guaranteed either the health or death of their specimens—alerted them to the consequences of unchecked imperial expansion and the need for ecological action.

Victorian orchid literature also complicates the boundaries we have both erected and removed in our own disciplinary considerations of earlier eras. In many respects, Victorians conceived of orchids not as colonizable "others" but as companionate species that share many human features and attributes—and that affect humans with both physiological and affective force. Such conceptions enable the imagination of what anthropologist Anna Tsing, echoing Bennett, has described as a form of "distributed agency" that focuses on the "entwined relations of humans and other species."[55] By questioning the necessary linkage of agency with intention, Tsing's notion unsettles the logic that privileges human subjectivities and that separates them from variably constructed "others," including botanic others. Both Tsing and Bennett, as well as Chen, loosen and even dissolve the boundaries that have promoted the view of plants as radically discontinuous with human ontology. Without such boundaries, concepts of community and social life are altered as well. Bennett theorizes a form of "political ecology" that would acknowledge the corelationships of humans with nonhuman entities and re-form itself as social circumstances and problems shifted. "If human culture is inextricably enmeshed with vibrant, nonhuman agencies," she writes, "and if human intentionality can be agentic only if accompanied by a vast entourage of nonhumans, then it seems that the appropriate unit of analysis for democratic theory is neither the individual human nor an exclusively human collective but the (ontologically heterogeneous) 'public' coalescing around a problem."[56] In her view, a sense of public responsibility would be intensified rather than attenuated by this form of "political ecology" because it builds on a strengthened sense of human identification with nonhuman entities.

Most Victorian horticulturists would not share Bennett's conclusions, for even as they identified with their orchids, they also maintained a firm sense of their own exceptionalism. That sense of exceptionalism, however, sometimes prompted a corresponding sense of responsibility that may well have been motivated by the awareness that their ecological fortunes were intertwined with those of orchids and other botanic species, both exotic and native. For Kate Soper, the very concept of human exceptionalism addresses the looser, fuzzier versions of posthumanist thought without blunting its political force. "To point out that we are all inter-connected in 'nature' and share much more with other animals [and plants] than we previously thought is all very well," she observes. "But what is important *eco-politically* is recognition both of the role of humans in bringing about ecological collapse, and of the distinctive capacities humans alone have to monitor, and in principle, to adjust their behaviour and environmental impact."[57] Clearly, Soper and Bennett would part ways at certain points in (post)humanist arguments. Yet their discord serves to highlight the rich potential of such thought and shows how Victorian ecological advances anticipated and even prepared the ground for such arguments. For Victorian horticulturists attuned to their orchids, humans were indeed exceptional—in their capacity both to destroy and to sustain fragile species and environments. If their ecological theory was not yet fully formed, its taxonomic, ontological, and epistemological ambiguity complicates our current views of them and even addresses our own ideologies and political commitments. What makes Victorians and their orchids seem so very strange is precisely what makes them significant today.

Notes

1. H. G. Wells, "The Strange Orchid," in *Thirty Strange Stories* (New York: Harper and Brothers, 1898), 11, 14.

2. Of Edward Said's many works that employ this paradigm, see in particular *Orientalism* (New York: Random House, 1978).

3. Jane Bennett, *Vibrant Matter: A Political Ecology of Things* (Durham, NC: Duke University Press, 2010), 21.

4. Unless otherwise indicated, the term "exotic" in this essay is used to indicate a nonnative species.

5. Lewis Castle, *Orchids: Their Structure, History, and Culture* (London: "Journal of Horticulture" Office, [1866]), 40.

6. James Bateman, *The Orchidaceae of Mexico and Guatemala* ([London: James Ridgway, 1837–54]), 3.

7. Lindley published several important volumes on orchids, a lifelong interest of his, and provided crucial impetus for orchid culture in nineteenth-century Britain. See [John] Lindley, *Folia Orchidaceae: An Enumeration of the Known Species of Orchids*, vol. 1 (London: Published for the author, by J. Matthews, 1852–55); John Lindley, *The Genera and Species of Orchidaceous Plants* (London: Ridgways, 1830–40); John Lindley, *Sertum Orchidaceum: A Wreath of the Most Beautiful Orchidaceous Flowers* (London: James Ridgway and Sons, 1838). Other botanists were instrumental as well in furthering knowledge and cultivation of exotic orchids in Britain. See, for example, J. D. Hooker, "A Century of Indian Orchids Selected from Drawings in the Herbarium of the Botanic Garden, Calcutta," *Annals of the Royal Botanic Garden, Calcutta* 5, no. 1 (1895): 1–16; Bartle Grant, *The Orchids of Burma (Including the Adaman Islands)* (Rangoon: Hanthawaddy Press, 1895); G. King and R. Pantling, *New Orchids from Sikkim* (Calcutta: Baptist Missions Press, 1895).

8. *Curtis's Botanical Magazine, or Flower-Garden Displayed* (London: [various publishers], 1801–1920); Conrad Loddiges and Sons, *Botanical Cabinet* (London: John and Arthur Arch, 1817–33); Robert Warner and Benjamin Samuel Williams, eds., *Orchid Album* (London: B. S. Williams, 1882–97); *Orchid Review* (London: West, Newman, 1893–1923); *Gardeners' Chronicle and Agricultural Gazette* (London, 1844–73); *Gardeners' Chronicle: A Weekly Illustrated Journal of Horticulture and Allied Subjects* (London, 1874–1922); "*Phalaenopsis Rosea,*" illustration in *Curtis's Botanical Magazine* 2 (1860): 5212.

9. Londa Schiebinger, *Plants and Empire: Colonial Bioprospecting in the Atlantic World* (Cambridge: Harvard University Press, 2004), 195. See also Richard Drayton, *Nature's Government: Science, Imperial Britain, and the "Improvement" of the World* (New Haven: Yale University Press, 2000) and Jim Endersby, *Imperial Nature: Joseph Hooker and the Practices of Victorian Science* (Chicago: University of Chicago Press, 2008).

10. Albert Millican, *Travels and Adventures of an Orchid Hunter: An Account of Canoe and Camp Life in Colombia, While Collecting Orchids in the Northern Andes* (London: Cassell, 1891), 149–51.

11. Ernst Haeckel, *Generelle Morphologie der Organismen: Allgemeine Grundzüge der organischen Formen-Wissenschaft, mechanisch begründet durch die von Charles Darwin reformirte Descendenz-Theorie,* vol. 2 (Berlin: Reimer, 1866), 286, quoted in Robert C. Stauffer, "Haeckel, Darwin, and Ecology," *Quarterly Review of Biology* 32, no. 3 (1957): 140. The quoted section from Haeckel was translated by Stauffer.

12. Stauffer, "Haeckel, Darwin, and Ecology," 139–40, 143. See also Frank N. Egerton, *Roots of Ecology: Antiquity to Haeckel* (Berkeley: University of California Press, 2012), 80–84, 198–200.

13. Charles Darwin, *On the Origin of Species by Means of Natural Selection,* ed. Joseph Carroll (Peterborough, ON: Broadview Press, 2003); Charles Darwin, *The Various Contrivances by Which Orchids Are Fertilised by Insects,* 2nd ed. (London: John Murray, 1890).

14. "A Newspaper Correspondent on Orchids." *Orchid Review* 4, no. 7 (January 1896): 203.

15. F. W. Burbidge, *Gardens of the Sun: Or a Naturalist's Journal on the Mountains and in the Forests and Swamps of Borneo and the Sulu Archipelago* (London: John Murray, 1880), 212; Ashmore Russan and Frederick Boyle, *The Orchid Seekers: A Story of Adventure in Borneo* (London: Frederick Warne, n.d.), 160; Frederick Boyle, *The Woodlands Orchids Described and Illustrated, with Stories of Orchid Collecting* (London: Macmillan, 1901), 246–47; Millican, *Travels and Adventures,* 118.

16. Alexander von Humboldt, *Personal Narrative of Travels to the Equinoctial Regions of the New Continent, during the Years 1799–1804,* 2nd ed., trans. Helen Maria Williams, vol. 2 (London: Longman, 1818). For an analysis of Humboldt's response to the tropical forest, see Lynn Voskuil, "Sotherton and the Geography of Empire: The Landscapes of Mansfield Park." *Studies in Romanticism* 53, no. 4 (2014): 591–615.

17. Boyle, *Woodlands Orchids,* 253.

18. Frederick Boyle, *About Orchids: A Chat* (London: Chapman and Hall, 1893), 86.

19. Darwin, *Various Contrivances,* 102.

20. Darwin, *Origin of Species,* 364.

21. Bateman, *Orchidaceae of Mexico,* 6.

22. Boyle, *About Orchids,* 206.

23. Lindley, *Sertum,* n.p.

24. Ibid.

25. Ibid.

26. "*Cypripedium Lowii*," illustration in Thomas Appleby, *The Orchid Manual, for the Cultivation of Stove, Greenhouse, and Hardy Orchids* (London: Journal of Horticulture and Cottage Gardner Office, 1861), 51.

27. Mrs. Talbot Clifton, *Pilgrims to the Isles of Penance: Orchid Gathering in the East* (London: John Long, 1911), 306.

28. Geo[rge] Hansen, *The Orchid Hybrids. Enumeration and Classification of All Hybrids of Orchids Published up to October 15, 1895* (London: Dulau, 1895), 48–49.

29. Cary Wolfe, *What Is Posthumanism?* (Minneapolis: University of Minnesota Press, 2010), xxix.

30. Paul Outka, "Posthuman/Postnatural: Ecocriticism and the Sublime in Mary Shelley's *Frankenstein*," in *Environmental Criticism for the Twenty-First Century*, ed. Stephanie LeMenager, Teresa Shewry, and Ken Hiltner (New York: Routledge, 2011), 31, 32.

31. For an excellent analysis of Darwin's work with orchids, see Devin Griffiths, "Flattening the World: Natural Theology and the Ecology of Darwin's Orchids," *Nineteenth-Century Contexts* 37, no. 5 (2015): 431–52.

32. Darwin, *Various Contrivances*, 163, 165.

33. Ibid., 23, 25.

34. James H. Veitch, *Hortus Veitchii: A History of the Rise and Progress of the Nurseries of Messrs. James Veitch and Sons* (London: James Veitch & Sons, 1906), 99.

35. Boyle, *About Orchids*, 233–234.

36. Darwin, *Various Contrivances*, 180.

37. Justin Prystash, however, has demonstrated the role of Darwin's eroticized botanical writing in feminist thinking as well. See "Fertilizing Darwin's Flowers: Feminist Narratives of Evolutionary Botany," *VIJ: Victorians Institute Journal* 39 (2011): 227–57.

38. W. Lauder Lindsay, "Mind in Plants," *British Journal of Psychiatry* 21, no. 96 (1876): 521, 525; Charles Darwin, *On the Movements and Habits of Climbing Plants* (London: Longman, Green, Longman, Roberts and Green, 1865).

39. Lindsay, "Mind in Plants," 523.

40. W. H. White, *The Book of Orchids* (London: John Lane, The Bodley Head, 1902).

41. [Donald] Beaton, [no source], quoted in Castle, *Orchids*, 44.

42. Mel Y. Chen, *Animacies: Biopolitics, Racial Mattering, and Queer Affect* (Durham, NC: Duke University Press, 2012), 4.

43. Bennett, *Vibrant Matter*, 120.

44. Boyle, *About Orchids*, 64.

45. Ibid., 71–72.

46. Ibid., 112.

47. Boyle's ecological leanings are not unique even among colonialists and imperialists. As Richard H. Grove has discussed and amply documented, environmental thinking and practice was often spurred by the experience of Europeans living and working in various colonial and global outposts. See Grove, *Green Imperialism: Colonial Expansion, Tropical Island Edens and the Origins of Environmentalism, 1600–1860* (Cambridge: Cambridge University Press, 1995) and Grove, "Conserving Eden: The (European) East India Companies and Their Environmental Policies on St. Helena, Mauritius and in Western India, 1660–1854," *Comparative Studies in Society and History* 35, no. 2 (1993): 318–51.

48. Boyle, *About Orchids,* 112; *Woodlands Orchids,* 145.

49. See, for example, James Britten and W. H. Gower, *Orchids for Amateurs: Containing Description of Orchids Suited to the Requirements of the Amateur* (London: "The Country" Office, [1878]); F. W. Burbidge, *Cool Orchids, and How to Grow Them* (London: Robert Hardwicke, 1874); Benjamin Samuel Williams, *The Orchid-Grower's Manual,* 4th ed. (London: Victoria and Paradise Nurseries, 1871.

50. W[illiam] Watson, *Orchids: Their Culture and Management,* 2nd ed. (London: L. Upcott Gill, 1903), 21.

51. Boyle, *About Orchids,* 224–25.

52. Ibid., 228.

53. Eric D. Brenner et al., "Plant Neurobiology: An Integrated View of Plant Signaling," *Trends in Plant Science* 11, no. 8 (2006): 414, 417. See also Michael Pollan, "The Intelligent Plant: Scientists Debate a New Way of Understanding Flora," *New Yorker,* December 23, 2013.

54. Lucinda Cole et al., "Speciesism, Identity Politics, and Ecocriticism: A Conversation with Humanists and Posthumanists." *Eighteenth Century* 52, no. 1 (2011): 96–97.

55. Anna Lowenhaupt Tsing, "Strathern beyond the Human: Testimony of a Spore," *Theory Culture Society* 31, nos. 2–3 (2014): 223, 224.

56. Bennett, *Vibrant Matter,* 108.

57. Kate Soper, "The Humanism in Posthumanism," *Comparative Critical Studies* 9, no. 3 (2012): 367.

CHAPTER 2

Discriminating the "Minuter Beauties of Nature"

Botany as Natural Theology in a Victorian Medical School

Meegan Kennedy

> The man of the world . . . will find a knowledge of botany [to be] a new source of pleasure.
> —Edward Forbes, *An Inaugural Lecture on Botany*

In his 1843 pamphlet *An Inaugural Lecture on Botany, Considered as a Science, and as a Branch of Medical Education,* the naturalist Edward Forbes argues that botany, rather than anatomy or physiology, should anchor the medical curriculum. Forbes, professor of botany at King's College, had himself left medical school as a youth to devote his life to natural history. In a series of rhetorical switchbacks, his lecture oddly promotes both botany's practical relevance—its relevance to pharmacology and to reasoning skills—and its pleasures, beauties, and even fancies. This unusual emphasis on pleasure allows Forbes to argue that botany functioned as an antidote to medical study, providing spiritual discipline and bringing morality into the corrupting atmosphere of the teaching hospital. Such a claim would not have been uncontroversial. Given botany's grounding in the contested sexual categorizations framed by Carl Linnaeus, its status as either science or moral compass was dubious. Forbes, however, positions botany as a purifying scientific activity by aligning it with what Barbara Gates calls the "narrative of natural theology," which considers the study of natural history to be spiritually and morally uplifting.[1] However, Forbes is speaking at a time when that narrative was being superseded in medicine by an empirically based clinical curriculum. His lecture demonstrates that, even as its medical faculty explored new laboratory sciences like physiology and chemistry, King's College privileged an older, natural theology model wherein medical and moral paradigms explicitly reinforced each another. Forbes strategically used his institutional context to negotiate a unique solution to the pedagogical and professional demands of the moment. He offers an understanding of botany as a dual pursuit: both a modern science that trains the student in the pragmatic skills of clinical observation and reasoning, and a traditional skill that hones the moral judgment and aesthetic appreciation of the man within. Forbes presents a fascinating counterpoint to the pioneering John Lindley, a vocal member of the anti-Linnaean, modernizing force in botany, who taught at the University of London. Although Forbes also taught using the system promoted by reformers, he contends that botanists should still revere the Linnaean system despite its limitations. He draws upon the narrative of natural theology to argue that botany's moral (old school) value in fact produces its (new school) scientific and intellectual value. Forbes's text demonstrates the careful negotiations between competing values and contexts that were important in professionalizing both botany and medicine.

Forbes wrote at a moment when scientific workers were—as Richard Yeo has shown—still debating the nature and meaning of science, either by drawing upon the authority of natural theology, or by attempting to establish a culture distinct from it.[2] This turn toward method (that is, the development of established procedures for scientific investigation) changed the meaning of virtue for the man of science; Forbes's lecture offers a balance between earlier and later, explicit and implicit models of scientific virtue. Steven Shapin locates the ideals of "Stoic fortitude and self-denial . . . disengagement and integrity" in eighteenth-century science, replaced in the late-Victorian period by the specialized expertise and technocracy of the methodical scientist.[3] Lorraine Daston and Peter Galison offer a different model, identifying the wisdom of the sage in the eighteenth-century model of objectivity they call "truth to nature," which was supplanted by an ideal of self-denial, exactitude, patience, and disinterested inquiry in the nineteenth-century model they call "mechanical objectivity."[4] Both Shapin's and Daston and Galison's models posit a movement from an explicit, external expression of scientific virtue to an implicit one. The economy of natural theology underlying eighteenth- and early nineteenth-century science was widely thought to produce an explicit virtue, a godliness, in its students, as its ultimate goal. Shapin attributes this to what was considered to be the sublime subject matter of God's Creation, the uniquely revelatory quality of natural knowledge, and the character of the "priests of nature" who chose to study Creation.[5] For Daston and Galison, the "mechanical objectivity" of nineteenth-century science (which can overlap with natural theology) then relies upon and produces a practical virtue implicitly, as a kind of by-product or side effect, through long-standing habits of laborious scientific work. For Shapin, twentieth-century technocracy, which he expressly opposes to natural theology, also relies upon implicit virtues like familiarity and reliability in its practitioners.

Forbes's lecture, however, delicately embraces *both* botany's history in natural theology and its newer, more scientific and professional systematics; and he cannily suggests that botany produces a hybrid bloom: both the older and newer, the spiritual and practical, the explicit and implicit forms of scientific virtue. The careful balance of his argument becomes clear upon comparison with an earlier lecture by Lindley, who had mounted a wholehearted defense of botany as a modern science. Forbes and Lindley provide a useful case study of how a modern, impersonal model of science—achieved through rigorous attention to method and producing an implicit model of practical scientific virtue—developed unevenly, alongside a continued investment in a science explicitly grounded in personal morality and in the tradition of natural theology.[6] At the intersection between these two models, where Forbes's lecture stands, botany takes on not only a practical and a moral but also, as this chapter will show, a remarkably fanciful role.

Unsettled Questions

When Forbes gave his lecture, the medical curriculum was in a state of flux. Thomas Neville Bonner argues that "no question in the nineteenth century was more agitated than the precise formulation of the educational and practice requirements for becoming a doctor."[7] Education could range from a classical university education followed by medical training, to studies at a practical medical school, to a simple apprenticeship.[8] Physiology and chemistry were still young, and many students subscribed to extracurricular private lectures or gained scarce clinical experience through private courses.[9] By 1840 most medical curricula included anatomy with dissection, physiology (usually book-taught), *materia medica* (pharmacy), and practical clinical work; botany and natural history might be included but were not widespread.[10]

Early nineteenth-century botany was in a similarly tenuous state, so that Forbes's claims for its value in the medical curriculum were weakened by the associations of botany with theoretical discord, sexual immorality, popular science, and (paradoxically) genteel womanhood. Botanists were fundamentally riven by divisions over the work of Carl Linnaeus, who had formalized the study of plants in his *Systema Naturae* (1735) in a taxonomy that sought to comprehend all plants over the range of the globe.[11] Stressing classification and nomenclature, he rejected earlier methods categorizing plants based on flower color and shape (number and shape of petals, leaves, roots) or other features. Having proven that plants reproduce sexually, Linnaeus established a binomial nomenclature: a plant's stamens (male structures) indicated its class, and its pistils (female) its order. His pioneering binary system allowed names to be used simply for identification rather than description. By selecting for a predetermined characteristic, Linnaeus promoted an "artificial" system—arbitrarily prioritizing one structure, thought essential to all plants, above other attributes of a plant. "Natural" (materialist) systems were less discriminating, some even attempting total description; but a universal, limited (artificial) taxonomy made it easier for botanists to navigate the storehouse of known information about the vast variety of plants.[12] The restricted focus of the Linnaean system enabled its strengths: it was simple to understand and use; it was consistent and stable; it required only observation of (usually) easily visible features; it enabled fast, accurate reference to specific plant species; and it could be widely applied across different types of plants, with good predictive power.

But early nineteenth-century botanists noted drawbacks to the Linnaean system. It was rigid and limited, and it was difficult to apply the system to plants with nonessential sexual characteristics. Thus, even as it became increasingly influential, critics developed alternative systems. For instance, Antoine-Laurent de Jussieu, in his *Genera Planterum* (1789), followed his uncle Bernard in examining multiple affinities of a plant, including its environment.[13] Theirs was a comprehensive, "natural" system that was not focused on a single feature or even on morphology.

A. G. Morton argues that, as evolutionary thinking began to appear in the mid-eighteenth century, researchers increasingly sought more information about species, moving away from artificial models to natural ones. The emphasis in botany, he claims, "shifted from observation of a limited number of characters regarded as essential for identification and classification, to the investigation of species in the round with the aim of attaining as full as description as possible in terms of morphology and anatomy . . . [with] physiology, development and relations to the environment."[14] By the early nineteenth century, most academic botanists considered the Linnaean system to be wrongheaded, but the binomial scheme nevertheless persisted for its simplicity and consistency.

These disagreements threw the field into confusion. By 1799 the Linnaean Robert Thornton noted fifty-two competing botanical systems, a state of affairs that one botanist called "system-madness" and "epidemical."[15] In 1829, Lindley remarked that "in Botany the fundamental principles are still unsettled; the world is much divided about them, and the purpose of the science, except as an accomplishment, is far from being generally understood."[16] Even given that "science" was still a developing category, the uncertainties within botany hindered its claim to scientific legitimacy.

Botany was additionally burdened by controversy over the Linnaean "Sexual System" of plant classification. Linnaeus's focus on the sexual structures of plants and his frequent use of analogies between plant reproduction and human marriage suggested, to some, an unhealthy and even immoral emphasis on lower functions—especially because few plant species are monogamous.[17] Erasmus Darwin's promotion of the analogy between human and plant sexuality and Sir Joseph Banks's reputation as a libertine underscored botany's dubious reputation, arousing restrictions on women naturalists and critique from conservative naturalists such as country clergymen.[18]

Botany also suffered from its association with the popular pursuit of natural history. As Lynn L. Merrill defines it, "natural history" is characterized by an interest in singularity (in the object for itself), the distinction between objects of study (as in a collection), and a "personal, evocative," and emotional tone. "Science," in contrast, is characterized by an interest in general laws (in the object for what it can tell us), the relationship between objects of study (again, as in a generalized law), and a neutral, "detached [and] objective" tone.[19] Nonetheless, naturalists could make important contributions to science; Merrill shows that the role of the gentlemanly amateur naturalist was entangled with that of the (at this time still rare) professional researcher in biology, botany, zoology, and geology. But the popularity of natural history led observers to underestimate its seriousness. Lindley says defensively, "Many people think that Natural History is nothing more than the amusement of shuffling or cutting natural objects, according to the caprices . . . of different observers."[20] Such perspectives were reinforced by phenomena like the "fern craze," which popularized the study of plants as a fad in interior decoration.[21]

Another aspect of botany that complicated its claims to scientific legitimacy was the view that it was primarily a genteel pastime for women. Despite plants' analogous relationship to human sexuality, they were thought to be more appropriate objects for women to study than land or sea creatures. When studied using sanitized texts that omitted reference to the crucial sexual structures, plants, bolstered by the narrative of natural theology, were considered spiritually uplifting and appropriate for women and the children they taught.[22] Frequently, botany was linked with the most domesticated, even feminized branch of natural history.[23] As Lindley complained, "It has been very much the fashion of late years . . . to undervalue the importance of this science, and to consider it an amusement for ladies rather than an occupation for the serious thoughts of man."[24] Many natural history texts for women, however, were written by professional researchers like Lindley. Also, women naturalists significantly contributed to the development of botany as a science. Mary Anning, famed for discovering the first *Icthyosaurus* fossil at age eleven, donated many botanic specimens to the Museum of Natural History. The American botanist Mary Treat collected for Asa Gray at Harvard and corresponded regularly with Darwin, correcting him on the traps of *Utricularia* (bladderworts) when her microscopic research disclosed tiny hairs triggering the trap to open.[25] Yet even the title of Treat's popular text, *Home Studies in Nature* (1880), reinscribes natural history as a private (domestic) rather than public (scientific) study.

Whether due to its theoretical uncertainties, its unseemly fascination with sexual structures, or its association with popular natural history and women's leisure, many questioned botany's legitimacy as a proper science. Consequently, botany's place in a medical education was questioned, particularly because accepted medical curricula focused on subjects thought to be practically useful. University of London instructors had even met resistance in the 1830s after proposing that the curriculum include theory of science, but medical schools eventually accepted the sciences of pathological anatomy, physiology, and to some extent, chemistry.[26] Botany, however, was not seen as being as immediately applicable to the practical or theoretical concerns of medicine. Forbes admits, "That the medical student acquires but little by his attendance at botanical lectures, is not an uncommon fancy among the senior members of the profession. Some eminent men have gone so far as to denounce it as lost time."[27] Indeed, Forbes followed the practice, common in Edinburgh, of teaching botany out of doors in the country,[28] but *Punch* had mocked just such excursions two years earlier as yielding only "chickweed, chamomiles, and dandelions," concluding, "The knowledge of the natural class and order of a buttercup must be of the greatest service to a practitioner in after-life in treating a case of typhus fever or ruptured blood-vessel."[29] In short, Forbes declared his dedication to botanical science at a time when its legitimacy and relevance to medicine was ever more in question.

"Men Must Be Educated into Such"

Despite this popular and academic resistance, Forbes maintained that botany is indeed practical for medical students. His *Inaugural Lecture on Botany*, however, also complicates the notion of what exactly is practical for medical students, suggesting that what is most immediately practical may not be what is, in the long run, the most valuable to a medical man. He agrees that botany is both scientific—"of all the natural-history sciences BOTANY is the most advanced"—and medical—it "forms a connecting link between professional and purely scientific studies."[30] If botany forms a bridge between science and medical practice, the most obvious argument for studying botany might be its pharmacological applications. However, *materia medica* was generally taught separately; perhaps for this reason, when Forbes does acknowledge the immediate practical value of botany in the pharmacopeia, he does so belatedly, reluctantly, and in passing, and as having only a secondary benefit. He says, "Though . . . the greatest benefit his botanical studies confer on the medical student is the making him a correct observer and careful reasoner, there is a fact-knowledge . . . of the greatest consequence in his profession. . . . [M]ore than 300 species of plants . . . furnish . . . articles of Materia Medica."[31]

However, Forbes emphasizes mental and moral uses over practical ones, praising botany's role in improving mind and character. He looks to longer-term goals, arguing for botany's ability to develop the empirical habits of mind that characterize the scientific physician. Botany, he claims, "train[s] the mind" to a state of "tone and vigour" by teaching the two most crucial skills of a physician, "*correct observation* and *accurate discrimination.*"[32] He further explains,

> The first depends mainly on the power of seizing all the features of an object or case with clearness and facility, detecting adventitious characters at sight, and excluding such from all influence on our conclusions. The second implies powers of just comparison, of perceiving the mutual relations of parts or facts, and of testing the possible agreement of statements with the circumstances which accompany them. Now though all men are endowed with the elements of these qualities, all are not born correct observers or accurate discriminators. Men must be educated into such. The mind must be trained to reason justly, the instruments of the mind to observe correctly.[33]

Here, by identifying fundamental elements of medical reasoning and asserting that they can be taught, Forbes aligns himself with aspects of the new clinical medicine, which similarly emphasized observation and methodology. Also, his emphasis on discrimination reflects the growing importance of nosology, the classification and diagnosis of disease according to its observable symptoms.[34] The distinguished pathologist Sir James Paget also links these practices, saying,

> I think it impossible to estimate too highly the influence of the study of botany. . . . It introduced me into the society of studious and observant men; it gave me an ambition for success . . . it encouraged the habit of observing, of really looking at things and learning the value of exact descriptions; it educated me in habits of orderly arrangement. . . . [The] unfelt power of observing and of recording facts . . . may justly be ascribed to the pursuit of botany. . . . [O]f the mere knowledge gained in the study . . . none had in my afterlife any measure of what is called practical utility. The knowledge was useless: the discipline of acquiring it was beyond all price.[35]

Forbes's claim that "the mind must be trained" and Paget's praise of "the unfelt power of observing and of recording facts" reflects a new emphasis in the 1840s on professional medical training. The Medical Reform Act regulating professional training and standards did not pass until 1858, but sixteen similar bills had been proposed from 1840. Since the 1830s, physicians had shifted away from book learning and ad hoc apprenticeship toward formalized programs in metropolitan schools of medicine.[36] Contemporary medical treatises and journals, like Thomas Wakley's reformist *Lancet*, demonstrate the influence of empirical methods and a new emphasis on the skills of clinical observation and reporting.

In his pamphlet, Forbes also outlines how botany practically promotes these desirable qualities of "correct observation and accurate discrimination." He posits,

> The first lesson of natural history is observation. The study of an animal or vegetable species is the perfection of observation. . . . The study of a group or genus of animals or vegetables is in like manner the perfection of discrimination. . . . The mental process is the same at the bed-side of the patient and in the cabinet of the naturalist: its first element, correct observation, leading to correct diagnosis; the second, accurate discrimination, leading to sound methods of treatment.[37]

Forbes's interest here in the relation between individual and group categories aligns diagnosis and botanical classification. However, in both passages above, Forbes's examples of observing and discriminating overlap: "observing" includes "detecting adventitious characters at sight, and excluding" them from consideration. In contrast, the mechanical objectivity identified by Daston and Galison aims at a complete, unfiltered, and unmediated record, one without judgment. Forbes's construct is closer to the earlier paradigm that Daston and Galison identify with eighteenth-century science, "truth to nature." In that model, a sage examines natural objects vis-à-vis an ideal of that object, filtering out unwanted artifacts of the individual or accidental, just as Forbes insists that the student must practice judgment about what is worthy to observe. He endorses the new empirical skills of medicine but has not fully embraced the skepticism of mechanical objectivity, because he does not reject the work of discrimination that anchors the eighteenth-century practice of "truth to nature."

The inconsistencies of Forbes's argument culminate in a paradoxical conclusion. He implies that botanical study is valuable to medicine mostly insofar as it *lacks* practical value.[38] Indeed, he says students should learn observation and discrimination away from the bedside: "No training is so strengthening as that which separates the process from the object of the process." Forbes warns of the "great evil in medical education" of teaching content and skill together, insofar as it "leads to habits of loose reasoning, and blunts the most valuable power of detecting fallacies," and results in "professional works notoriously abound[ing] in bad logic."[39] With botany, however, "Who can rise up from such a study and not feel mentally strengthened?"[40] Forbes's eccentric reasoning comes to the fore here in his claims of the practical value of an impractical study; and in his critique of the Continental method of bedside clinical instruction, in favor of a supposedly more logical, if bloodless, training in medical skill by way of plant biology, precisely because botany has so little to do with the human body.

These peculiarities of Forbes's argument make sense, however, when he shifts to a third, implicitly greater, point: medical students should learn botany because its abstract reasoning will make them not just better doctors but better men. His rhetoric references an older tradition of science that suggests botany improves not only the mind but also character. This moral formation could have had a practical application, since British medical students were notorious for drunken hijinks, violent outbursts, and coarse or ribald talk.[41] John Stevens, in a Scottish midcentury tract attacking man-midwifery, warns,

> Imagine, forty or fifty gay young fellows, full of midnight sprees, and half-and-half, surrounding the patient with those sacred parts of her person all exposed to their indecent gaze! Bright and peering is the eye-glass of the dissipated fop, insinuating is the gross jeer of the medical libertine, the habitual scoffer at all human virtue.[42]

Contemporaries suspected that medical study itself elicited this rowdy, coarse behavior, rendering hospitals dangerous for patients' physical and students' moral health. John James Audubon, after visiting the dissecting theater of surgeon Robert Knox (later infamous for his role in the Burke and Hare case), recalled, "I was glad to leave this charnel house and breathe again the salubrious atmosphere of the streets."[43] Both the dissections and the discourse were offensive; even the professors used bawdy language and told obscene stories in the dissection room and the lecture hall.[44]

Forbes was well aware of the unsavory reputations of both hospitals and their students, so he prescribes a particular form of learning to guard against the miasma of the hospital:

> One great evil which has tended to retard the intellectual advancement of the medical student, especially in this great city, has been

> the separation of his studies from all association with the pursuits of the scholar and the philosopher. The air of a hospital is mentally unwholesome, unless mingled with a full proportion of collegiate atmosphere.

Part of the problem was simply the narrowness of hospital study, focused exclusively on the body; Forbes warns that an "exclusive professional education" causes a dangerous "contraction of the mind." But he also hints at immorality in terming hospital air "unwholesome," in contrast to "the very neighbourhood of literary and scientific studies," which "has a purifying and elevating effect on the mind of the student."[45] Botany—as both intellectual exercise and natural theology—epitomizes this spiritual discipline, providing a "new source of pleasure" to jaded medical students, who thus acquire wholesome "after-occupations" away from the hospital.[46] Because botany is not directly relevant to hospital work, it is not defiled by that environment and can correct for the taint of medical study. Here again, Forbes offers a series of counterintuitive claims: that botany is productive to medical students because it is irrelevant to them; that it is useful for their occupation because it is enjoyable as an "after-occupation"; that it is morally purifying because it is so pleasurable. But these help explain Forbes's puzzling shifts. While the historical context of medical curriculum reform demanded that he tout the practical utility of botanical study, he believed botany valuable for its blend of practical and moral skills—that is, for its ability to train the skills needed in medicine but also to *distance* students from the coarse atmosphere of the hospital and dissecting room.

Forbes's argument about botany's moral and practical utility puts into focus the tensions that characterized shifting beliefs in contemporary medical thought. For instance, he draws on the miasma theory of disease in describing the corrupting nature of hospital air. Early nineteenth-century physicians proposed climatological theories of illness, which emphasized the dangerous effects of environment more than the individual vector of contagion. Certain locales were thought to be miasmatic, where the bad air carried disease, and temperature and wind direction might influence health by rendering bodies more vulnerable to illness. In her 1860 book *Notes on Nursing*, Florence Nightingale argues that "the first essential to a patient" is "to keep the air he breathes as pure as the external air," and she exhorts her readers to "always air your room, then, from the outside air."[47] Just as fresh air provided a wholesome, health-giving element in blowing away the stale air of the sickroom, studying open-air plants, Forbes suggests, promotes mental and moral health by opening the narrow, pernicious confines of medical study to the fresh air of botanizing. Forbes's vision of the virtuous medical man also shares little with the bourgeois discipline demonstrated by the emergent practitioner of mechanical objectivity, who strives to be patient, focused, and capable of heroic efforts of nearly mechanical reportage. Critical of this paradigm, Forbes claims that, "shut out from the spirit of letters, of science and of art, exclusively occupied with one set of thoughts and practices, the man sinks into the drudge." But botanical study makes "the young physician and surgeon . . . a scholar, a man of science, and a man of taste; and, above all, imbued with sound principles of religion and morality."[48] Forbes's use of "scholar" here nods to the classical education still common for physicians. His mention of "science" acknowledges the rise of clinical medicine; while his interest in "taste" recalls the crucial role of discernment and suggests that medicine should remain a profession in the older sense, marked by gentility more than education, certification, and society membership.[49] Forbes places religion and morality above the classical foundations of medicine, its increasing claims to science, or its traditional status as a profession.

Forbes's argument that botany improves moral health follows logically from botany's roots in natural history, especially the narrative of natural theology: the philosophy, popularized by William Paley's 1802 text *Natural theology, or, Evidences of the existence and attributes of the Deity*, that natural history is the record of Creation and that a close reading of that record can produce a wonder engendering a better knowledge of God.[50] Peter Mark Roget had explained, in his *Animal and Vegetable Physiology* (1834), the fifth Bridgewater Treatise arguing for the trace of God in the Creation: "To Man have been revealed the Power, the Wisdom, and the Goodness of God, through the medium of the Book of Nature." The student of Nature, by "contemplation" of natural objects, feels "admiration and . . . gratitude" and "refines" his soul.[51] Minuteness, beauty, mechanical contrivance, wonder, and the sublime ground the narrative of natural theology; the Paleyite observer first marvels at the tiny structures of nature, miniscule but apparently perfect, then experiences a spiritual epiphany. Roget says of the fibrils of feathers,

> A construction so refined and artificial . . . and so perfectly adapted to [its] mechanical object . . . cannot be contemplated without the

> deepest feeling of admiration, and without the most eager curiosity to gain an insight into . . . such minute and curious workmanship. . . . [N]one is more fitted to call forth our profoundest wonder at the comprehensiveness of the vast scheme of divine providence.[52]

Forbes engages this narrative to argue that botany became more potent a tool for theological conversion as it became more scientific. Studying plants, he says,

> led at length from empiricism to science . . . the herborist ripened into the botanist [and a] new light broke upon him. . . . The wonders of [plant] structure were exposed. . . . [T]heir history became a store, from whence could be drawn at pleasure numberless admirable examples of the perfection of design in creation, and of the benevolence and omniscience of the Creator.[53]

In fact, Forbes argues that natural theology actually inspired botanical science, when "earnest unbiassed studies originating in the admiration of the wonders and beauties of creation, and deep reverence for the great Origin of all things, were the corner-stones of botanical science."[54] Even in his own day, he claims, those "who have journeyed much in foreign lands have felt the delight of examining some beautiful and strange flower . . . and many an idler has been metamorphosed into a man of science by . . . such accidental direction of his attention to the minuter beauties of nature."[55] Many contemporary botanical texts tapped the narrative of natural theology like this; for example, in the journal *Botanist,* Samuel Maund cites Paley as inspiration.[56] But most of these write for a general audience, exemplary of the growing divide Ann B. Shteir identifies between moral ("polite") and scientific botany. In contrast, Forbes addresses a professional audience and argues that botany has value for that audience precisely because of its links to natural theology; that natural theology makes botany more, not less, scientific. He thus strives to reintegrate polite botany with scientific botany even as he adopts the new, natural systematics. If, as I've been arguing, Forbes acknowledges the faultlines of botanists' split between an "aesthetic, moral, and spiritual orientation . . . [and a] utilitarian or scientific approach,"[57] he also contends that, by reintegrating the strands of botany, naturalists might secure it greater cultural authority as a foundation of the new medical science as well as the established moral tradition.

The Clue to the Labyrinth

Forbes's unwillingness to relinquish botany's past treasures—whether natural theology or Linnaean taxonomy—becomes clear in comparison to a similar lecture delivered fourteen years earlier. The reformer John Lindley, like Forbes, had been a new botany professor publishing his inaugural remarks to the University of London.[58] Most of Lindley's examples, unlike those of Forbes, are not moralistic but resolutely pragmatic: we use vegetable matter in many daily activities, so we should study plant life;[59] the physician relies upon plant-based medicaments, so he should study plants and their properties.[60] Lindley also lauds horticulturists like Thomas Andrew Knight, who achieved "the complete subjugation of the unmanageable constitution of the Pine-apple."[61] Artists, too, need botany for practical reasons: botanical study prevents embarrassing "blunders" such as painting "flowers stuck upon parts where they could not more have grown than a man's head beneath his arms."[62] Although Lindley offers an extended, lush description of plant life, his aim is determinedly utilitarian: "If the vegetable world is thus indispensable to our very existence, and if it is really subject to the influence of certain fixed laws, can it be doubted that it is of the utmost importance to the world to be acquainted with these laws? And what is that acquaintance but Botany?"[63]

It may seem counterintuitive that Lindley, writing earlier, endorses a practical scientific approach to botanic study, while Forbes, the later writer, ultimately endorses both botanical science and a moralist approach informed by botany's traditional link with natural theology. Lindley pressed for the newer, "natural" system of classification of Jussieu, which acknowledges multiple aspects of the plant, to replace the "artificial" Linnaean system, which considers only stamens and pistils. Both Lindley and Forbes taught a Jussieuian system—Forbes had actually joined Jussieu's class briefly while visiting Paris[64]—but their emphasis in discussing the predecessor Linnaeus is telling. Lindley damns the artificial system for "rendering Botany a mere science of names, than which nothing more useless can be well conceived."[65] He concludes that the Linnaean system is "a positive and serious evil."[66] In her reading of the text, Shteir recalls the association between the Linnaean system and botany for ladies. Although Linnaean terms were sometimes considered inappropriate for women, his system was the default for traditionalist and popular texts. Thus Lindley's "rejection of Linnaean botany is a rejection of polite botany in favor of utilitarian botany [and by] the mid-1840s . . . literary botany and scientific botany became distinct discourses."[67]

In contrast, Forbes, nearly fifteen years later, urges botanical science to make peace with its roots in natural theology and to embrace the moral strengths promised by its contemporary, genteel associations. Forbes rows against the tide of increasing science, professionalism, and specialization here, offering a hybrid vision of old and new, polite and scientific botany. His embrace of the "minuter beauties" of botany and natural theology apparently inspires him to return to Linnaeus with similarly romantic prose. He lavishes two pages on Linnaeus, praising him for his "two ingenious artificial schemes": "a universal [botanical] language" that offered "the greatest means of furthering the progress of natural history," and "the making of an index to a great section of the book of nature."[68] The shift away from science and method is evident in Forbes's language. When he needs a rationale for keeping the Linnaean system, critiqued by so many others, Forbes uncharacteristically turns to a far-fetched, fanciful metaphor. The Linnaean system is "a most valuable auxiliary," he says, because

> the understanding of things depends greatly on the perception of their order and relations. When that order and those relations require deep study . . . the man who gives us a clue, however insignificant it may be in its own nature, . . . endow[s] the despised instrument with golden value. Such a clue did Linnaeus give when he put forth the sexual system. . . . The clue to the labyrinth, then, having served such noble purpose becomes a consecrated object, and should rather be hung up in the temple than thrown aside with ignominy. The traveler returning from his adventurous and perilous journey of discovery, hangs up his knapsack with affection on the wall of his study.[69]

Just as the science of botany leads us to recognize the wonders of creation, Forbes says, the Linnaean system led us to recognize the value of taxonomy. Here he reframes the rationalist "order and relations" of taxonomy as a romantic tale. The botanist becomes a seeker in an unknown land, questing in a labyrinth, carrying a noble, consecrated object (the Linnaean system) that provides a clue to those mysteries. This excursion metaphorically transports us to an exotic, dangerous locale. A generally favorable notice of Forbes's lecture in the *Annals and Magazine of Natural History* wryly comments, "We are quite willing to hang this system up in the temple anywhere as long as it does not interfere with plants . . . we should never wish again to disturb its dignity by carrying it into the fields."[70] But Forbes insists these flights of fancy, including aspects of the Linnaean system itself, are compatible with modern scientific realism. Just as British botanists ventured into colonial lands for specimens to bring home for study and reference, Forbes implies, Linnaeus led botanists through the mysteries of the plant world until—Forbes deftly returns us to the quotidian—the homecoming botanist hangs the battered knapsack (the no-longer needed sexual system) on the "wall of his study." Forbes acknowledges that the system was considered "insignificant," but he also describes it in terms of wonder and mystery, a mode familiar from natural theology. He acknowledges that the sexual system is "despised" but urges us not to cast it away. Rather, we should recognize it as the mystical "clue to the labyrinth," a "consecrated object" with "golden value."

Given Forbes's emphasis on rational processes of observation and the general shifts at the time toward a more scientific medical curriculum, his fanciful, elaborate metaphor may seem a peculiar rhetorical choice. However, this metaphor of the labyrinth, and its sacralization of Linnaean polite botany as the clue to nature's mysteries, allows Forbes to pivot back to moral concerns. This shift signals both his position within the debates over curriculum and his institutional context. Lindley had spoken to, and from, the University of London, whereas Forbes addressed his students and colleagues at King's College. The University of London, also then known as "London University," was founded as an alternative to Oxford and Cambridge in 1826. The founders, inspired by the radical educational and social philosophy of Jeremy Bentham, had Jewish, utilitarian, and dissenting support in establishing a secular university for the urban middle classes. The university offered courses in new areas of study like political economy, English literature, classics, and science. In 1828, King's College was founded as a religious, Anglican, and traditionalist response to the University of London, but because neither of the two competing institutions had degree-granting authority, they combined as constituent colleges ("University College" and King's College) of the newly chartered University of London in 1836 to secure this right. These years were marked by acute competition between the schools, and, as J. Reynolds Green explains, "Botany was one of the chairs affected. Lindley was made Professor at University College in 1829, and King's, not to be outdone, founded a Chair only two years later."[71]

Lindley, then, aligned his aims with those of his secular university in promoting rationalist investigation and the new sciences: "In this new Institution we will see . . . whether it is not possible to found a school of Botany in London worthy of being associated with those of Medicine, Zoology, and Natural Philosophy."[72] Forbes also associates his aims with those of his college, but in a very different institutional context and in very different terms:

> That which Lord Bacon said of all knowledge is especially true of this department, that it "is not a couch whereupon to rest a searching and restless spirit; or a terrace for a wandering and variable mind to walk up and down with a fair prospect; or a tower of state for a proud mind to raise itself upon; or a fort or commanding ground for strife and contention; or a shop for profit or sale;—but a rich storehouse for the glory of the Creator and the relief of man's estate."[73]

Such a conclusion neatly dismisses other motivations as crassly materialistic and reinstates the narrative of natural theology in asserting that knowledge—here, the study of botany—is valuable simply for its insight into the products of a divine Creator. Forbes's defense of Linnaeus aligns with this allegiance to botany's idealized origin as moral and spiritual instructor. Although Forbes endorses Jussieu earlier in this text, his conclusion—and its embrace of metaphor—promotes a compromise between traditional and modern, fanciful and pragmatic, polite and systematic approaches to botany and the spiritual and practical virtues of science.

Forbes's attempt to blend polite and scientific botany did not reflect trends in the larger botanical community, but, surprisingly, he may have helped preserve a place for botany in the medical curriculum for a time. The *Annals and Magazine of Natural History* reviewed his remarks with approval, taking both Forbes and the new sciences seriously; in a nearby article, Arthur Hill Hassall responds to Forbes's earlier critique of his work on polypes.[74] In 1854, Forbes attained his life's goal, the appointment as professor of natural history at the University of Edinburgh (just before his untimely death). He reiterated his views on medical education in his inaugural lecture, and the acceptability of those views is evident in that the lecture was published in both the Edinburgh *Monthly Journal of Medical Science* and the *Medical Examiner*, even being quoted at length in a letter to *Nature* in 1883.[75] By 1855 botany was required for licensure as a surgeon or apothecary, and by 1884 required for physicians. Anxiety over the moral health of students persisted; the president of the British

Medical Association in 1868, William Stokes, voiced this concern in his address to the membership.[76] Botany was, however, eventually crowded out of the medical curriculum as the new sciences and practical clinical experience became necessary; indeed, T. H. Huxley singled it out (with zoology) as unnecessary to the medical curriculum in his 1874 lecture "Universities: Actual and Ideal."[77]

Forbes's and Lindley's texts, however, demonstrate the importance of institutional context and the uneven pace of scientific change. The University of London accommodated Lindley's condemnation of Linnaeus and his advocacy for the natural systematics advancing botanical science. The setting of the religious and traditional King's College allowed Forbes, years later, to champion botany on broader terms, echoing the moral narrative of natural theology. The struggle between the "art" and "science" of medicine was fervent, and botany was battling for professional recognition. But for Forbes the greatest value botany offered to students was not its contributions to the science of medicine—its pharmacological resources or ability to hone medical observation and discrimination—but its links to the art of medicine. By teaching the appreciation of minute beauties, Forbes believed, botany preserved this gentlemanly tradition and its moral strengths. His lecture reminds us that modernizing the medical curriculum involved not just innovation but also hesitation, negotiation, and compromise; and that medicine was, like other sciences, deeply involved in the changing status of natural history and natural theology in the nineteenth century.

Notes

1. Barbara T. Gates, *Kindred Nature: Victorian and Edwardian Women Embrace the Living World* (Chicago: University of Chicago Press, 1998), 38–39. Gates cites Greg Myers for the "narrative of natural history" and the "narrative of science." Myers, *Writing Biology: Texts in the Social Construction of Scientific Knowledge* (Madison: University of Wisconsin Press, 1990), 142–43.

2. Richard Yeo, *Defining Science: William Whewell, Natural Knowledge, and Public Debate in Early Victorian Britain* (Cambridge: Cambridge University Press, 1993), 29–35.

3. Steven Shapin, *The Scientific Life: A Moral History of a Late Modern Vocation* (Chicago: University of Chicago Press, 2008), 36–42.

4. Lorraine Daston and Peter Galison, *Objectivity* (New York: Zone, 2007).

5. Shapin, *The Scientific Life,* 82–83.

6. Indeed, Steven Shapin argues that personal virtue remains tied to "the scientific life" even into the era of contemporary technoscience; personal qualities remain

both central to science and under erasure in narratives about "what scientists do." Ibid., 4–5.

7. Thomas Neville Bonner, *Becoming a Physician: Medical Education in Britain, France, Germany, and the United States, 1750–1945* (New York: Oxford University Press, 1995), 63.

8. M. Jeanne Peterson, *The Medical Profession in Mid-Victorian London* (Berkeley: University of California Press, 1978), 5.

9. Bonner, *Becoming a Physician,* 81–85, 90–91, 100–101. For an American example, see Ephraim Cutter, *Partial Syllabic Lists of the Clinical Morphologies of the Blood . . .* (Trenton: [n.p.], 1892).

10. Bonner, *Becoming a Physician,* 145.

11. See Bettina Dietz, "Contribution and Co-production: The Collaborative Culture of Linnaean Botany," *Annals of Science* 69, no. 4 (2012): 551–69.

12. Carl Linnaeus proposed his ideas in his *System Naturae,* first published in 1735, the tenth (1758) edition of which set forth his binomial nomenclature. Linnaeus, *System Naturae* (Stockholm: L. Salvius, 1758). As Alan Morton points out, however, Linnaeus frequently used empirical evidence, basing his genera on natural groupings. Morton, *History of Botanical Science: An Account of the Development of Botany from Ancient Times to the Present Day* (London: Academic Press, 1981), 264, 298–308.

13. Antoine-Laurent de Jussieu, *Genera Planterum* (Paris: Herissant and Barrois, 1789). Bernard worked with Antoine-Laurent in the preparation of this work.

14. Morton, *History of Botanical Science,* 293.

15. Thomas Martyn, *Memoirs of John Martyn . . . and of Thomas Martyn . . . Professors of Botany* (Piccadilly [London]: Hatchard & Son, 1830), 135, quoted in David Allen, *The Naturalist in Britain: A Social History* (Princeton, NJ: Princeton University Press, 1994), 34, quoted in Londa Schiebinger, *Nature's Body: Gender in the Making of Modern Science* (Boston: Beacon, 1993), 14.

16. John Lindley, *An Introductory Lecture Delivered in the University of London . . .* (London: Taylor, 1829), [3].

17. See, for example, Amy King, *Bloom: The Botanical Vernacular in the English Novel* (Oxford: Oxford University Press, 2003), 19–29 and 50–58; Schiebinger, *Nature's Body,* 12–14, 28–37.

18. Samantha George, *Botany, Sexuality, and Women's Writing, 1760–1830: From Modest Shoot to Forward Plant* (Manchester: Manchester University Press, 2007). Banks was a renowned naturalist and instrumental in the work at Kew Gardens, but his exploits in Tahiti prompted gossip and satires on his supposed libertinage; see, for example Pamela Cheek, *Sexual Antipodes: Enlightenment, Globalization, and the Placing of Sex* (Stanford: Stanford University Press, 2003), 143–45, and Patty O'Brien, *The Pacific Muse: Exotic Femininity and the Colonial Pacific* (Seattle: University of Washington Press, 2006), 63–67. For the sexual system as immoral, see also O'Brien, *Pacific Muse,* 13–15.

19. Lynn Merrill, *The Romance of Victorian Natural History* (Oxford: Oxford University Press, 1989), 90–96.

20. Lindley, *Introductory Lecture,* 12.

21. For a description of the "fern craze," see David Elliston Allen, "Tastes and Crazes," in *Cultures of Natural History,* ed. N. Jardine, J. A. Secord, and E. C. Spary (Cambridge: Cambridge University Press, 1996), 394–407.

22. Gates, *Kindred Nature,* 38–39. See also Londa Schiebinger, *The Mind Has No Sex? Women in the Origins of Modern Science* (Cambridge: Harvard University Press), 243; and King, *Bloom,* 57.

23. For the feminine associations of botany, see Gates, *Kindred Nature*, 64, 140; Ann B. Shteir, "Women in the Polite Culture of Botany," in *Cultivating Women, Cultivating Science: Flora's Daughters and Botany in England, 1760–1860* (Baltimore: Johns Hopkins University Press, 1996), 33–58; Schiebinger, *Mind Has No Sex*, 241–44; Samantha George, "'Unveiling the Mysteries of Vegetation': Botany and the Feminine," in *Botany, Sexuality*. For attempts in the 1830s to masculinize botany and distinguish professional from popular botany, see Shteir, "Defeminizing the Budding Science of Botany," in *Cultivating Women, Cultivating Science*, 147–70.

24. Lindley, *Introductory Lecture*, 14.

25. Mary Treat, *Home Studies in Nature* (New York: Harper & Brothers, Franklin Square, 1885), 147–62, esp. 147, 157–60, 162. The Darwin Correspondence Project notes that Treat exchanged fifteen letters with Darwin, far more than most female naturalists represented. "The Public and Private Face of Mary Treat," DarwinLetters, posted by Philippa Hardman, accessed June 10, 2015, http://www.darwinproject.ac.uk/gender/2011/10/12/the-public-and-private-face-of-mary-treat.

26. Bonner, *Becoming a Physician*, 144, 156.

27. Edward Forbes, *An Inaugural Lecture on Botany, Considered as a Science, and as a Branch of Medical Education* . . . (London: J. van Voorst, [1843]), 7.

28. [John Hughes Bennett], "Biography of the Late Professor Edward Forbes," *Monthly Journal of Medical Science* 1 (January 1855): 82.

29. "The Physiology of the London Medical Student, No. 4: Of the Manner in Which the First Season Passes," *Punch*, October 23, 1841, 177.

30. Forbes, *Inaugural Lecture*, 6, 3.

31. Ibid., 12. Forbes lauds the practical use of botany in many nonmedical fields, including agriculture, chemistry, zoology, geology, and in a military or colonial context. "Even to the man of the world it may afford profit and pleasure," he says. Ibid., 13–15.

32. Ibid., 7, 8.

33. Ibid., 8.

34. Bonner, *Becoming a Physician*, 14; Charles Newman, *The Evolution of Medical Education in the Nineteenth Century* (Oxford: Oxford University Press, 1957), 97.

35. Stephen Paget, *Memoirs and Letters of Sir James Paget* (London: Longmans, Green, 1901), 27–28. Paget's memoirs were written in 1880–85 and published in 1901. Here he recalls his apprenticeship before hospital-pupillage, 1830–34.

36. For a good overview of this transition, see Bonner, *Becoming a Physician*.

37. Forbes, *Inaugural Lecture*, 9–10.

38. This is perhaps not surprising, given Forbes's lack of sympathy with medicine, a study he took up to please his father. John Hughes Bennett, a fellow student at Edinburgh, recalled upon Forbes's death that "he could never conquer his dislike to medicine as a profession. He was seldom seen in the dissecting-room or Infirmary. Even his attendance on the purely medical classes was of no great use to him, as he did little but sketch the features of the professor or of the surrounding students." "Biography," 76. Another biographical account notes, "As for medicine it had no interest for him, except as including certain departments of Natural History, and for its own more special studies he had an unconquerable aversion." George Wilson and Archibald Geike, *Memoir of Edward Forbes, F.R.S., Late Regius Professor of Natural History in the University of Edinburgh* (London: Macmillan, 1861), 70; see also 184–87.

39. Forbes, *Inaugural Lecture*, 10.

40. Ibid., 9.

41. Newman, *Evolution of Medical Education*, 41–47; Peterson, *Medical Profession*, 40; Bonner, *Becoming a Physician*, 8, 72–75, 215.

42. John Stevens, *Man-Midwifery Exposed*, 2nd ed. (London: William Horsell, n.d.), 16–17. In Scotland, half-and-half is a dram of whisky with a chaser of beer. Stevens's text is undated, but it is collated in the Wellcome Library in a collection titled "Aberdeen Medical-Chirurgical Society, Tracts (1850s–1871)."

43. Maria R. Audubon, *Audubon and His Journals*, vol. 1 (New York: Scribner's Sons, 1899).

44. Newman, *Evolution of Medical Education*, 42–45; Bonner, *Becoming a Physician*, 210.

45. Forbes, *Inaugural Lecture*, 11.

46. Ibid., 14, 8.

47. Florence Nightingale, *Notes on Nursing* (New York: Dover, 1969).

48. Forbes, *Inaugural Lecture*, 11–12.

49. Susan Faye Cannon, *Science in Culture: The Early Victorian Period* (New York: Dawson and Science History Publications, 1978), 145ff.

50. William Paley, *Natural Theology: Or, Evidence of the Existence and Attributes of the Deity* (London: R. Faulder, 1802).

51. Peter Mark Roget, *Animal and Vegetable Physiology Considered with Reference to Natural Theology*, vol. 1 [Bridgewater Treatise V] (London: William Pickering, 1834).

52. Forbes, *Inaugural Lecture*, 571–72.

53. Ibid., 6.

54. Ibid., 17.

55. Ibid., 14.

56. Samuel Maund, "Preface," *Botanist* 1 (1836), quoted in Shteir, *Cultivating Women, Cultivating Science*, 159.

57. Shteir, *Cultivating Women, Cultivating Science*, 153.

58. Lindley, *Introductory Lecture.*

59. Ibid., 17–18.

60. Ibid., 19–21.

61. Ibid., 22.

62. Ibid., 23.

63. Ibid., 18.

64. Wilson, *Memoir of Edward Forbes*, 219.

65. Lindley, *Introductory Lecture*, 9.

66. Ibid., 10.

67. Shteir, *Cultivating Women, Cultivating Science*, 157–158.

68. Forbes, *Inaugural Lecture*, 16–17.

69. Ibid., 19.

70. "Bibliographical Notices: An Inaugural Lecture on Botany . . . ," *Annals and Magazine of Natural History: Zoology, Botany, and Geology* 12, no. 75 (1843): 123.

71. J. Reynolds Green, *A History of Botany in the United Kingdom from the Earliest Times to the End of the 19th Century* (London: J. M. Dent, 1914), 363–64.

72. Lindley, *Introductory Lecture*, 14.

73. Forbes, *Inaugural Lecture*, 23. Forbes slightly misquotes Bacon here, who opens the passage, "But the greatest error of all the rest, is the mistaking or misplacing of the last or farthest end of knowledge . . . as if there were sought in knowledge a couch . . . and not a rich storehouse." Francis Bacon, *Of the Proficience and Advancement of Learning, Divine and Moral*, in *The Works of Lord Bacon: With an Introductory Essay, and a Portrait*, vol. 1 (London: William Ball, 1838), 13.

74. "Bibliograpical Notices," 120–23; Arthur Hill Hassall, "Observations on Two

of Professor Edward Forbes's 'Retrospective Comments,'" *Annals and Magazine of Natural History: Zoology, Botany, and Geology* 12, no. 75 (1843): 117–20.

75. Forbes explains, "It is this training of the mind in correct methods of observation that gives the Natural History sciences so much value as instruments of preparation in professional education." He demonstrates a familiar impatience with the insistence on practical subjects, exclaiming, "Not unfrequently do we hear the short-sighted and narrow-minded ask, what is the use of zoology, or botany, or geology to the physician and surgeon? what have they to do with beasts, or plants, or stones? Is not their work among men, healing the sick? . . . Vain and stupid questions all." And he offers a similar emphasis on "wholesome" food for the mind: "Were the sciences . . . to be entirely professional, we should warp and contract the minds. . . . [T]he collateral sciences [like botany] . . . prevent an undue dissipation of the student's thoughts" (563), "Inaugural Lecture. By Edward Forces, F.R.S., F.G.S., Professors of Natural History in the University of Edinburgh. Delivered March 15, 1854." *Monthly Journal of Medical Science* 18 (July 1854): 560–68. Republished in part as "Upon the Pleasures and Advantages of a Knowledge of the Natural Sciences," *Medical Examiner: A Monthly Record of Medical Science* 10 (July 1854): 437–39 and quoted at length in John H. Balfour, letter to the editor, *Nature*, December 18, 1873, 121–22.

76. Peterson, *Medical Profession*, 55.

77. For information on the medical curriculum, see Peterson, *Medical Profession*, 62; Bonner, *Becoming a Physician*, 217–18; Thomas Henry Huxley, "Universities: Actual and Ideal," in *Collected Essays*, vol. 3: *Science and Education* (London: Macmillan, 1893), 189–234.

CHAPTER 3

"A Perfect World of Wonders"

Marianne North and the Pleasures and Pursuits of Botany

Narin Hassan

In the preface to Marianne North's travel narrative, *Recollections of a Happy Life* (1894), Catherine Symonds writes: "My sister was no botanist in the technical sense of the term; her feeling for plants in their beautiful living personality was more like that which we all have for our human friends. She could never bear to see flowers uselessly gathered—their harmless lives destroyed."[1] Symonds highlights how North straddled cultures of science and art, and of collection and environmentalism; though not "technically" a botanist, North clearly loved, studied, and depicted plants on a massive scale. She traveled the world to explore and to paint specimens from tropical landscapes but, as her sister notes, preferred to see flowers blooming in their native habitats. She had relationships with some of the renowned nineteenth-century men of science—her narrative describes her encounters with Charles Darwin, Alfred Russell Wallace, Joseph Hooker, and many others—and, although she was not professionally trained as a scientist or artist, she was recognized for her discovery and depiction of rare plant species. North lived a privileged life quite unusual for Victorian women and formed friendships with travelers and writers including Amelia Edwards, Lucie Duff Gordon, and

Isabella Bird, and artists Julia Margaret Cameron and William Holman Hunt. She was one of the most active and intrepid female explorers and had the means to travel the entire globe while narrating her experiences and recording her visions in a robust collection of paintings and illustrations. She produced hundreds of visual images and donated 832 of her paintings to the Royal Botanic gardens at Kew, where she funded the development of the North Gallery and supervised the display of her works. "She had the lower walls lined with boards made of the 246 different types of wood from which she had collected on her travels." She also "paid for 2,000 copies of a catalog compiled at her request by . . . Kew botanist W. Botting Hemsley."[2] These images, like her travel narrative, record an active life of exploration and reveal North's unique vision of the globe and natural life within it.

Although, as Symonds notes, North could not bear seeing plants "uselessly gathered," her representations contributed to expanding archives of botanical knowledge and she hovered on the cultural edges of both scientific and imperial expansion. She defined a unique relationship between the woman traveler and landscape: in her narrative and paintings she highlights the power of plants and their potential relationship with humans as "friends." North's "feeling" for plants and emphasis upon her intimate immersion within natural habitats challenges assumptions about nineteenth-century botanical exploration, including notions of the prototypical Victorian woman traveler as a distanced observer of landscape and more interested in native peoples, domestic lifestyles, and customs. While the tropics were a space where North could feel emancipated from the confines of Victorian society and participate in discourses of both science and colonialism in ways she could not have at home, her unique representations of the environment provided alternative, and sometimes subversive, visions of tropical spaces. Although her depictions unveil the wonders and specimens of the natural world, they also emphasize human interaction with nature as an intimate and phenomenological experience. In an age when botanical culture was increasingly linked with classificatory systems and objective notions of scientific knowledge, North promotes a subjective and fluid vision of nature focused upon immersive feeling and sensory expansiveness.

Marianne North became a naturalist, painter, writer, traveler, and amateur botanist during a time when the "nature" of botany as a field was becoming professionalized. As Ann Shteir has shown, botany shifted from being a domestic form of amusement to becoming a more scientific and professional venture in the nineteenth century.[3] But it was also increasingly associated with expanding conceptualizations of scientific knowledge and imperial conquest that often disregarded the activities of women. While, on the one hand, botanical interactions were linked to domestic amusement for ladies, on the other, as Mary Louise Pratt has famously argued, botanical knowledge was tied to masculine scientific and imperial pursuits:

> Natural history called upon human intervention (intellectual, mainly) to compose an order. . . . One by one the planet's life forms were to be drawn out of the tangled threads of their life surroundings and rewoven into European-based patterns of global unity and order. The (lettered, male, European) eye that held the system could familiarize ("naturalize") new sites/sights immediately upon contact, by incorporating them into the language of the system of nature.[4]

David Arnold, in his reading of the tropics, ties the desire for order in European representations of nature to colonial histories: "Ideas of landscape, far from being peripheral to the exercise of power or merely reflective of a material reality, formed a central and integrating element in the wider constitution of colonial knowledge and a critical ingredient in the larger colonizing process."[5] Further, Janet Browne reminds us that "the study of animal and plant geography in the nineteenth century was one of the most obvious imperial sciences in an age of expanding imperialism."[6]

North challenges binary notions of botany as either a ladylike domestic pursuit or an expanding science associated with masculinity and imperial expansion during this age. Instead, she articulates the ways that women can engage with botany at a global scale and reconstitute notions of the exotic within the natural world. Instead of imposing a preconceived vision onto the landscapes she encounters, North engages with the process of viewing and perceiving foreign spaces as a subjective and evolving experience. She presents an archive of botanical description and illustration that contributes to scientific knowledge and colonial discourses while also emphasizing the sensational pleasures, the wonder, and the intimate enjoyment of landscapes as powerful and elusive sites of transformation. She represents nature as containing rich and complex processes that cannot be easily categorized or captured by human interests. Instead, for North, plants function as independent and interdependent systems that sometimes challenge scientific objectification and categorization. As an intrepid traveler, North provides us with a complex image of the Englishwoman abroad as an active, immersed, and embodied agent of scientific inquiry who participates within, and reframes, our sense of imperial science through her sustained commitment to representing nature; she is simultaneously engaged with, and ambivalent toward, the project of imperialism. North's focus upon the intimate qualities of her experience with the natural world and the sensations they produce open readers and viewers of her work to the expansiveness of global knowledge and the complex qualities of nature. In her travel writing, paintings produced overseas, and the North gallery she curated at Kew upon return to England, plants are given a majestic and powerful status, depicted not as victims or specimens of imperial expansion, but instead as powerful and appealing agents that can build human consciousness and create a sense of global connectivity.

"Vegetables Suited Me Better": Placing North within a Critical Context

North is a perplexing figure, and criticism of her work emphasizes that she, like the plants she describes, often defies categorization. In her introduction to *Recollections of a Happy Life,* and her book, *Place Matters: Gendered Geography in Victorian Women's Travel Books about Southeast Asia,* Susan Morgan describes the difficulty of defining North as artist, writer, and traveler. Reminding us that botany was itself an emerging science within this period, Morgan interrogates North's complex role as amateur scientist, and ponders her complicity within imperial science, examining "how scientific imperialism is being represented when the leading imperialist role is being filled by a female subject."[7] Like Morgan, Antonia Losano considers North's gendered position, arguing that her work "exhibits a rhetoric of emancipation."[8] These critics show that North moves beyond the ladylike pursuit of botany toward a more authoritative role as a gendered subject participating on the fringes of imperial science. Barbara Gates also describes the dilemma of the female explorer and the different expectations for women engaging in scientific discovery, reminding us that even though North led an active life of botanical discovery, she "still did not win the acclaim the scientific community routinely awarded to men who had accomplished far less than she."[9] More recently, building upon the work of Morgan and others, Eadaoin Agnew has suggested that North bolsters colonial projects in the development of her own professional and authoritative voice, claiming that "North actively endorses the scientific projects of nineteenth-century imperialism."[10]

North's visual imagery has also garnered attention, and a number of recent studies, such as Michelle Payne's *Marianne North: A Very Intrepid Painter* and Laura Ponsonby's *Abundant Beauty: The Adventurous Travels of Marianne North, Botanical Artist,* provide a stunning overview of her visual images.[11] Yet even in the area of artistic production North is hard to define. Whereas Ponsonby, as the subtitle to her volume suggests, claims North as a "botanical artist," Suzanne LeMay Sheffield describes North as one who did not quite fit into the traditions of either artistic or scientific illustration and claims the kind of work she produced was elusive and hard to define.[12] Michelle Payne describes her representations as "nature study," whereas Lynn Merrill designates her as a "scientific artist."[13] Karen Morin also notes that "it is difficult to place North as a botanist or a more scientific type of naturalist; it is also difficult to place her as an artist."[14] North is an elusive figure, shifting between nineteenth-century cultures of science, imperialism, art, and travel, but ultimately challenging the boundaries of these areas.

It is clear that North sought the acknowledgment of scientific men and claimed some authority through her discoveries and expansive journeys into tropical lands. Accordingly, she may be considered a conduit to imperial expansion and certainly may be assumed to align herself with Western science. She does, as Morgan and others have suggested, produce the "naturalist's gaze" by objectifying the beauty of plants and revealing the human desire to know and consume landscape.[15] I also agree that she does engage with and benefit from expanding systems of imperialism and scientific exploration. In the early pages of her narrative, she describes her visits to Chiswick gardens to paint "specimen flowers" and to Kew, where, she writes, William Hooker "gave me a hanging bunch of *Amherstia nobilis,* one of the grandest flowers in existence. It was the first that had bloomed in England and made me long more and more to see the tropics."[16] The very nature of nineteenth-century travel—particularly botanical exploration—was already deeply embedded within structures of imperial science during this time, and to "survey" the globe assumed a desire to define and consume it within cultures of imperial knowledge and expansion. North's competitive and ambitious desire to discover rare species, and the fact that five species of plants were named after her, reveals her implication within scientific imperialism.

North also, however, challenges traditional nineteenth-century practices of collection and questions the pursuits of plant hunters who tend to ignore the interconnected qualities of natural life, suggesting that plants sometimes "gaze back" and challenge our efforts to collect and contain them. Her textual narrative highlights the overwhelming powers of nature, and her visual images provide equally potent representations of plants as unique and perplexing figures that both interact with and challenge human intrusion. Indeed, at numerous moments within her text, North alludes to her own connection to plants as far more enticing than human interaction. In the second volume of her *Recollections*, she follows a short description of native women in Borneo exotically "clad in all the colours of the rainbow" with the claim, "Vegetables suited me better."[17] While North asserts her identity as a painter, traveler, and writer through her tropical pursuits, she also creates a holistic vision of landscape that emphasizes the power of plants and their ability to sometimes overwhelm and captivate humans. Her representations reshape conceptions of nature in an experimental, uncanny, and strange way—highlighting the sensory and immersive possibilities of human engagement with plant life. In creating her own perceptive and unique relationship with landscape, North acknowledges the inevitable progress of an imperial moment; but she also suggests that such projects may be challenged, as the natural world may have its own unruly, unpredictable, and uncontainable systems.

"Trespassing on Fairies' Grounds": Reading the Sensations and Representations of North's Travel Narrative

Contemporary scholarship on Victorian women travelers tends to hover between viewing these figures as either uniquely equipped to challenge masculine stereotypes of representation or caught within the rhetoric and machinery of imperialism as exhibited in their colonial representations and imagery. North's botanical pursuits implicate her within imperial expansion, but also allow her to retain and celebrate the unique natural environments in which she immersed herself. Her narrative shifts and complicates the structures of gender and science as it does notions of race and culture. Unlike many women's travel journals of the time, hers has few descriptions of social, personal, or cultural life within the landscape she experienced. Her *Recollections* mention a few of the British expatriates she encountered on her travels, and provide some descriptions of native habits and lifestyles, but her main focus is upon the variety of flora and fauna she encounters and the strategies she employs as an intrepid and careful plant seeker and adventuress.[18]

North's travel narrative departs from traditional tropes that dominated women's travel narratives and instead gives her readers a vision of vast landscapes and sublime scenes that emphasize the unattainable qualities of the natural world. Breaking the tradition of women's travel accounts that often focused upon domestic life, titillating scenes of harems, or encounters with native servants and cultural traditions, North instead provides a sweeping, picturesque display of larger-than-life geographical scenes—which, nonetheless, are exotic and sensual. When she does describe native servants, ayahs, and children, she suggests that they hinder the important work of illustration and painting. In the opening pages of volume 2 of her *Recollections* she writes: "Two ayahs followed me in, and fought for the possession of me, though I wanted neither."[19] Abandoning her role as domestic manager or bearer of respectable British values, North instead focuses upon the world outside the confines of English compounds and homes. During her time in Jamaica, she writes, "after about a month of perfect quiet and incessant painting . . . people began to find me out" and "I begged to let off formal breakfasts, went out after my cup of tea at sunrise as I did at home, and worked till noon."[20] In general, people are not her focus; they are the background to her paintings, while botanical subjects are in the foreground, and her narrative seems to function in a similar way. Once she arrives in Darjeeling, "the finest hill place in the world," she describes the sublime qualities of the landscape and her need for solitude to digest its wonders:

> The flowers about Darjeeling seemed endless. I found new ones every day. The *Thunbergia coccinea* was perhaps the most striking; it twined itself up to the tops of the oaks, and hung down in long tresses of brilliant color, the oak itself having leaves like the sweet chestnut, and great acorns as big as apricots almost hidden in their cups. There was another lovely creeper peculiar to Darjeeling,—the sweet-scented cluster ipomoea, of a pure pink or lilac color. The wild hydrangea with its tricolor blooms was also much more beautiful than the tame one. I worked so hard and walked so much that after a dinner or two with Sir Ashley Eden and other grandees, I refused any more invitations.[21]

In this description, North privileges the "wild" hydrangea over the tame one and presents her readers with a visual archive of the plants she encounters. The flowers are "endless" and varied and far more intriguing to her than the English inhabitants she encounters—she favors her interaction with plants over her "invitations" from others and depicts these as distracting from her work. North's narrative traces an endless stream of visits to botanical gardens and walks through forests to study plants and record them within their natural environments. She immerses herself completely within these tropical "pleasure gardens," viewing them microscopically and from all angles. In New South Wales she writes: "At the lake of Illawarra we again found ourselves in the tropics, all tangled with unknown plants and greenery, abundant stag's-horns, banksias, hakea, and odd things."[22] Both she, and the plants she describes, have physical qualities that become entangled and fused. From one location to another, she records the wonders of being submerged within nature and plant life—highlighting the joy of the rare discoveries with strong and spirited proclamations: "In the jungle, I found real pitcher-plants (Nepenthes) winding themselves amongst the tropical bracken. It was the first time I had seen them growing wild, and I screamed with delight."[23] The pleasure and delight of garden hunting far exceed the need for ladylike behavior when it comes to North's descriptions, and in one scene after another she describes what appears to be her unending joy in engaging with plants. North's descriptions focus upon the sensations she feels during her journey and inspire readers to imagine the density and intricacy of the wild tropics. She depicts herself as a figure embodied and submerged in the tropics and presents plants as her equal sensory partners within the landscape. But unlike popular images of the tropics as dangerous and potentially unhealthy or unappealing spaces, North presents them as attractions that can provide thrilling and transformative experiences.

Like other women travelers of the nineteenth century, North describes her discoveries as a source of liberation and self-awareness. In Chile, one object of her pursuit is the "blue puya" plant, a rare specimen that she became aware of before her travels. She is driven to see the plant and writes:

> Of course the first thing I tried to get was the great blue puya. I was told they were all out of flower; indeed, some people declared they did not exist, because they had not seen them. At last an energetic English lady bribed a man to bring me one from the mountain. It was a very bad specimen, but I screamed with delight at it, and worked hard to get it done before it was quite faded, for it was past its prime.[24]

Although a specimen is brought to her, North is determined to find the plant on her own and witness it within its natural habitat. After climbing up a steep mountain, she writes:

> I could not see a yard before me, but would not give up and was rewarded at last by the mists clearing, and behold, just over my head, a great group of the noble flowers, standing out like ghosts at first, then gradually coming out with their full beauty of color and form in every stage of growth; while beyond them glittered a snow peak far away, and I reached a new world of wonders, with blue sky overhead, and a mass of clouds like sheets of cotton-wool below me, hiding the valley I had left.[25]

North initially shows delight over the specimen she is presented with, but her real goal is to experience the plant's powerful aura in its natural environment. For North, these "noble" flowers represent a magical vision—both ghostly and wondrous. Her pursuit of plants is strangely scientific and supernatural at the same time; while she is well versed in the scientific nomenclature of plants and comfortably describes plants using their Latin names within her text, she also highlights a spiritual, sacred aspect to plant life that moves beyond objective definition. North's pilgrimage up this steep mountain is a somewhat treacherous but also contemplative experience, providing her with an almost mystical sense of the environment. Her visit allows her to witness new wonders rarely seen, and she exposes the experience to her readers, creating a distinctive vision in both her narrative account and visual sketches. Although North was known for her very secular beliefs, her descriptions of botanical subjects emphasize that plants are an entryway to the unknown and to the sacred. She produced a series of paintings on sacred plants of India, and described native beliefs about the hidden healing powers of plants. After discovering the puya plant in Chile she comments that "the gum of the plant is valuable in medicine," and in India she notes her work of "hunting up the Sacred plants" and learning about them from a "learned baboo" who said "it pleased him much that I should take so much trouble about the plants that Siva loved."[26]

North contrasts these wild, joyous, mystical, and immersive descriptions of her travels with scenes in London where she describes the fate of tropical plants on display. Recording a walk with Asa Gray and his wife in London, she writes:

> One day I went with them both through Veitch's hot-houses, and we were shown all his wonderful hybrid orchids, with the parent plants, and the clever man who hybridized them. We saw also houses full of pitcher-plants, baby pitchers, not bigger than pins' heads, including the "*Nepenthes Northiana*," in search of which a traveller had been sent across the world to Borneo after seeing my painting at Kensington. But it will be difficult to imitate, in a cramping glass house in foggy London, the abundance of air, though hot, in those limestone mountains.[27]

Although much of North's early engagement with plants was through the domestic glasshouses her father built, she privileges the experience of witnessing plants in their "natural" states over the process of propagating and transporting them in artificial surroundings.

While North may be read as an agent of imperial expansion, she also appears as an environmental activist arguing for the need to grow and retain plants in their native settings. As Dea Birkett has noted, North engaged with the larger network of scientific and colonial pursuits overseas and established several greenhouses by her home where she grew and painted tropical orchids and specimens.[28] But her narrative also describes the need to preserve natural environments and suggests that plants cannot thrive when uprooted from their tropical landscapes and transferred to the artificial conditions in hothouses. For example, in a passage that describes the joy of "hunting up all kinds of orchids" and hanging them to view, she writes: "I fear few of my treasures lived long."[29] In the highlands of Brazil, she describes nature as gorgeous, enticing, and yet unattainable: "Gorgeous flowers grew close, but just out of reach, and every now and then I caught sight of some tiny nest, hanging inside a sheltering and prickly screen of brambles. All these wonders seem to taunt us mortals for trespassing on fairies' grounds, and to tell us they were unapproachable."[30] In an age of imperial and scientific expansion, North alludes to the inevitable desire to hunt plants and the urge to consume or dominate tropical lands. And yet her descriptions also suggest a circular dynamic between plants and the world of imperial science—one in which the tropical gaze is returned to remind "more mortals" of the potentially challenging dynamics of trespassing.

As her narrative unfolds, North establishes the process of painting as completely immersive, consuming, and addictive—she is lost in the landscape and the image she is creating, and her descriptions suggest the power of plants to entice and attract humans. Day after day she describes the pleasures of her "work" and oil painting as an irresistible experience. Describing a scene upon her arrival in Jamaica, she writes: "I was in a state of ecstasy, and hardly knew what to paint first."[31] North's narrative emphasizes the sensational qualities of engaging with nature; while she documents the plants she sees in a thorough and systematic way, she is driven by feeling, and her framing of the natural world—in both her journal and her images—emphasizes the intimacy of her relationship with the plants she encounters. She describes the "never-ending delight" of gardens, and her memoir traces landscapes and specimens from one chapter to the next in an ever-expanding process of description and aesthetic immersion.[32] The highlands of Brazil are "a perfect fairy land" (1:118), and a fern walk is described as having "lovely, fairy like beauty" (1:89), the ferns being "most delicious to look at" (1:89). While her catalog of plant life makes botanic specimens an object of study and aesthetic vision, she seems to challenge the nineteenth-century pursuit of plants as objects of circulation and travel—opting instead to produce a unifying, global vision of plant life that connects one part of the globe to the other, and that produces a massive "wonderland" of plant life across the globe.[33] Thus, her narrative creates a catalog of sensational representations that entice readers to imagine, from a distance and up close, the wonders of the natural world.

North's Visual Perceptions: Image and Display

North's visual work also displays her immersive and experiential approach, both through her use of materials (primarily oil paint) and through the unique perspective, depth, and framing of her subject matter. Much of the botanical representation produced by women in the Victorian period was through sketching and watercolor—this was partly due to an established tradition of feminine styles of painting and the fact that women often painted at home and not in larger studios. As a number of critics, including Sheffield and Losano, have noted, North, however, chose to produce oil-based paintings, and she produced works that were aesthetically and spatially different from those of botanical illustrators of the time. Working with oils was a more complicated process—especially for an artist traveling to different locations—but oil paint captured the lushness and rich color of her surroundings and enhanced the visually stimulating and layered qualities of her images. With oil paint,

North could produce more vivid images that could capture the majestic qualities of plants that she described in her writing. Instead of floating independently on a page, her plant forms burst out of colorful scenes, and often her paintings capture one part of a larger scene—evoking a sense of the grandeur of the landscape. She is experimental with her paintings, and her vision is flexible and broad. Some of her paintings provide lush, panoramic scenes with an expansive focus, while others take a microscopic, albeit sensual and fleshy, approach. North describes the process of trying to capture the intensity and breadth of the natural world in her art:

> When I went to finish another sketch I was astounded at the sight of a huge lily, with white face and pink stalks and backs, resting its heavy head on the ground. It grew from a single-stemmed plant, with the grand, curved leaves above the flower, and was called there the Brookiana lily, but Kew magnates call it *Crinum augustum*; its head was two feet across, and I had to take a smaller specimen to paint in order to get it into my half-sheet of paper life-size. It was scented like vanilla. Another crinum has since been called Northiana, after myself. It has a magnificent flower, growing almost in the water, each plant becoming an island at high tide, with beautiful reflections under it, and its perfect white petals enriched by the bright pink stamens which hang over them.[34]

While in this case North finds a specimen to fit on her paper in life-size form, sometimes a portion of a leaf or plant will burst from the seam of a painting, or burst in from the edges evoking how plants exist in a large, rich, exotic landscape, not in isolation. As Patricia Murphy notes, North avoids the "subtle shades typically associated with Victorian woman's artwork." Instead, she "frequently opts for bold, startling colors that seem to issue a visual challenge to an observer."[35] The visual intensity of North's work—in her saturated and deep choice of colors, and direct, immersive choice of design—have inspired critics to describe her work as frightening and unsettling. Lynn Merrill writes: "The brilliant colors and otherworldly rococo forms of the plants in her tropical paintings pulsate with garish ominousness: they are so *other*, they are almost frightening. Botanical records, grounded in fact and precise observation, they nonetheless spark emotional fires."[36] Further, Antonia Losano notes that North "paints vegetable chaos. Flowers tumble over the canvas, one scientifically interesting part of one flower hidden by another. . . . One always has the sense of something—animal, mineral, vegetable, spiritual—lurking in the shadows at the boundaries of the vividly cheerful flowers."[37]

North's images do seem to inspire a sense of otherness—one that I would suggest vindicates nature as a powerful force that is almost beyond representation. Instead of representing the "specimen-like" quality of a plant—its ability to float independently on a page or lie static in glass jar—her plants are consistently alive in their natural environments. North's paintings depict plants in a way that emphasizes their movement and growth, as well as their potential to overwhelm us in a sensory way. North seems acutely aware of the simultaneous vulnerability of nature and the power of it. Her paintings allude to cultures of scientific illustration, but are not concerned with providing simplified, objective renderings of exotic specimens. Instead, they provide a more immersive view of the natural world and inspire viewers to imagine being and feeling within a foreign landscape. One of many examples of this is her *Red Water Lily of Southern India.*[38]

North presents not one red lily, but many red lilies, and she depicts them at various stages of development and in tandem with other natural scenery. Like her description of the "great group of noble flowers" in Chile, she depicts the red lilies as a continually evolving and intertwined group of flowers. While the image has a close-up, almost microscopic view of the botanical subjects, it evokes the fullness of the landscape—there are plants on each side of the central lily, and several plants confront the viewer from the borders of the scene. The image includes insects that are hovering among the flowers, revealing the variety of natural life within the scene. North also depicts the image in a way that suggests the painter is directly engaged with the space—immersed within the plants and conveying the overall sensations they evoke. North confronts nature directly and refashions its representation, challenging and expanding our notions of Victorian botanical illustration and suggesting that while nature may capture the gaze of eager viewers, it also is constantly in flux. North catalogs plant after plant, in both her textual descriptions and her brilliant paintings, in a highly sensory way, and as she traces and records these visions from her travels, she provides a connective thread from one location to the next. Her vision for capturing the connections among natural life across the globe extends as she continues her travels from one continent to the next and she imagines capturing her global immersion visually and with a tactile approach.

Although North is remembered for her unusual and evocative imag-

Fig. 3.1. Marianne North, *Red Water Lily of Southern India,* 1878. (Courtesy of the Kew Royal Botanical Gardens.)

es and her entertaining narrative, she also produced her own unusual plant displays and recorded memories of her travels by creating physical arrangements and exhibits. When she travels to the United States she notes:

> I had intended on putting an enlarged map of the world on the ceiling, coloured according to the geographical distribution of plants, in different shades of green and brown, the sea also shaded as it is in nature. . . . I meant to add an index of fruits painted by myself, on the cornice, and twelve typical trees between the windows, but every one was against such an unconventional idea, except my old friend Mr. Fergusson, and he wanted some good geographer to make a model, and suggested consulting Francis Galton or Mr. Wallace. The first was most kind and helpful as usual, but covered the map he started on with level lines and curves from 500 to 10,000 feet, and that was of no use on so small a scale. Then I made a pilgrimage to see Mr. Wallace, and found him most delightful, and much interested in my plan. He recommended asking Mr. Trelawney Saunders to make my map, which he did,—a most exquisite piece of hand-shading for which I

> paid £120,—but it was not in the least what I wanted. . . . I also got woods from all parts of the world to make a dado of. . . . It was a great difficulty to arrange them, but time mended all. The catalogue I wrote on cards, and stuck them under the paintings; and after I had put down all I knew, Mr. Hemsley corrected and added more information, which he did so thoroughly and carefully that I asked him to finish the whole, and to put his name to the publication.[39]

I provide this lengthy quotation to emphasize how North creates her own unique process of "systemizing" nature in a sensory way. She has the social rank and experience to build relationships with these scientific men, but the attention she receives is limited as she succumbs to "Mr. Hemsley" to "put his name on the publication." She also has an unusual vision for how a project that displays nature could function and offers a varying approach to the cataloging and display of botanical culture. Her painted "index of fruits" is viewed as unconventional by the men who view it, although it represents the range of plants she has encountered. While she attempts to engage with the scientific trend of cataloging and displaying the "wonders" of nature, her approach contrasts with traditional methods of categorizing global plant life and instead attempts to capture feelings and sensations as she displays the rare plant specimens of the globe. She emphasizes the "curves" and the range of shades in her map, which functions not as a two-dimensional arrangement, but instead as a tactile and physical object with materials like wood from different parts of the globe. Once again, North reveals her interest in producing a global connectivity through plants while capturing and exhibiting nature. While she aligns herself with the men who were producing similar kinds of objects of scientific categorization, she also lets her readers know about her own unique and contrasting approach to botanical display as her "unconventional" ideas are dismissed. As imperial science worked toward compartmentalizing, hybridizing, and breeding nature, North focuses upon retaining and capturing its intrinsic systems while highlighting its wildness and diversity.

North's organization of her gallery at Kew Gardens provides a similarly unconventional mode of display. It is here that North appears to have had more of a say in the presentation of her work. North selected how her paintings should be hung, and within the gallery her paintings are displayed in very close proximity to one another as a massive collection of frames from floor to ceiling, creating, once again, an almost claustrophobic, but also immersive, sense of plant life around the globe.

Fig. 3.2. The Marianne North Gallery at Kew Gardens. (Courtesy of the Kew Royal Botanical Gardens.)

While the compressed display may be a result of the limited space within the gallery, it also supports North's global vision—her paintings are arranged in relation to the continents in which they reside, and the walls of wood from native trees connect the paintings to their original environments. While her exhibit produces what could be read as a vast imperial catalog of botanical life, it defies simple categorization and instead produces a more immersive and infused sense of the linkages between various geographies and environments. By placing the images directly next to one another, with hardly any space in between, North also produces a sense of the plants being linked together in an intimate way as they would within a natural environment.[40] Further, within the enclosed space of the gallery, the abundance of North's images, placed in a continuum, reproduce the overwhelming colors and sensations she would have witnessed in the tropics.

While, on the one hand, North's categorization and pursuit of plants opens them up to scientific inquiry and contributes to nineteenth-century colonial exploration, her display and representation of the plant world as an overwhelming, powerful, and somewhat secretive space provides an alternative vision of the natural world that challenges notions of scientific objectivity and order. In her narrative, paintings, and gallery,

North ultimately asks us to shift the expectations and boundaries of scientific and imperial knowledge and to expand our notions of scale and perspective. As her own narrative suggests, she herself did not want to be categorized in any one way. After a meeting with Louis Agassiz and his wife she writes, "Mrs. Agassiz and I agreed that the greatest pleasure we knew was to see new and wonderful countries, and the only rival to that was the one of staying quietly at home. Only ignorant fools think that because one likes sugar one cannot like salt; those people are only capable of one idea and never try experiments."[41] North certainly lived a life that pushed boundaries and was driven by a constant sense of the pleasures and possibilities of experimentation. Her sense of wonder allowed her to engage with imperial science in a way that brought to light the vulnerabilities and complexities of it—by producing powerful and uniquely exhibited scenes of nature, North asked viewers to to analyze the structures of botanical knowledge and to question what it meant to be a human subject assuming authority over natural life. She straddled the worlds of professional and amateur science, imperial culture and travel, and art and literature, constantly identifying herself with more than one discourse, and emphasizing the importance of spontaneity and feeling. Knowledge and subjective pleasure come together in North's "strange science," and within her narratives, she balances imaginative and subjective discovery and professional pursuit, producing a unique, perplexing, and visually stunning display of her travels that ultimately reveals the immense longing and bountiful pleasure North sought from nature.

Notes

1. Catherine Symonds, preface to *Recollections of a Happy Life*, by Marianne North, ed. Mrs. John Addington Symonds, 2 vols. (London: Macmillan, 1894), n.p.

2. Susan Morgan, introduction to *Recollections of a Happy Life*, by Marianne North, ed. Susan Morgan, 2 vols. (Charlottesville: University Press of Virginia, 1993), xii. See also her chapter on North in *Place Matters: Gendered Geography in Victorian Women's Travel Books about Southeast Asia* (New Brunswick, NJ: Rutgers University Press, 1996), 100–101, for more detail about the North's life and her gallery at Kew Gardens. Michelle Payne also provides biographical background about North as well as information about the formation of the North gallery. Michelle Payne, *Marianne North: A Very Intrepid Painter* (Richmond: Kew Publishing, 2011), 87–89.

3. See Ann Shteir, *Cultivating Women, Cultivating Science: Flora's Daughters and Botany in England, 1760–1860* (Baltimore: Johns Hopkins University Press, 1999). Shteir notes: "From the 1780s on, botany accorded with conventional ideas about women's nature and 'natural' roles. It fit the gendered assumptions of many writers

and arbiters about women and domestic ideology and was thought to be a way to shape women, or to shape them better, for their lives as wives and mothers" (35). She argues that during 1830–60 the direction of botanical culture became more "scientific" and professionalized with the influence of men such as John Lindley (153). Thus, botany and the study of the natural world was a growing academic practice as well as an entertaining and educational domestic pursuit. Barbara T. Gates has also traced the gendered shifts in the study of botany and natural history and the different tropes necessary for women in her book *Kindred Nature: Victorian and Edwardian Women Embrace the Living World* (Chicago: University of Chicago Press, 1999). For additional material on gender and botany, see Sam George and Alison E. Martin, eds., "Women and Botany," special issue, *Journal of Literature and Science* 4, no. 1 (2011).

4. Mary Louise Pratt, *Imperial Eyes: Travel Writing and Transculturation* (New York: Routledge, 1992), 31. Pratt outlines the "systematizing of nature" through many examples, including the important scientific expeditions of the seventeenth and eighteenth centuries and the development of Linnaeus's *Science of Nature* in 1735. Thus, she reveals how cultures of travel and botany both participated in the formation of a more systematic and classificatory way of viewing the globe.

5. David Arnold, *The Tropics and the Traveling Gaze: India, Landscape, and Science, 1800–1856* (Seattle: University of Washington Press, 2006), 5.

6. Janet Browne, "Biogeography and Empire," in *Cultures of Natural History: From Curiosity to Crisis*, ed. Nicholas Jardine, James Secord, and Emma Spary (Cambridge: Cambridge University Press, 1996), 305.

7. Morgan, *Place Matters*, 119. In her rich and thorough reading, Morgan concludes that North represents the "union of imperialism, women's emancipation, and the natural sciences in nineteenth-century Britain" (132).

8. Antonia Losano, "A Preference for Vegetables: The Travel Writings and Botanical Art of Marianne North," *Women's Studies* 26 (1997): 428.

9. Barbara Gates describes this dilemma of the female explorer: "A different set of tropes was necessary for women who had to work to make their scientific discoveries known. North had collected thousands of plants for Kew gardens and had discovered five species later named for her—Kniphofia northiana, Areca northiana, Crinum northianum, Nepenthes northiana, Northea seychellana, a tree of Seychelles (possibly the one she sketched for Allman), which Hooker named in her honor—but still did not win acclaim the scientific community routinely awarded to men who had accomplished far less than she." Gates, *Kindred Nature*, 99.

10. Eadaoin Agnew, "'An Old Vagabond': Science and Sexuality in Marianne North's Representations of India," *Nineteenth Century Gender Studies* 7, no. 2 (2011), http://www.ncgsjournal.com/issue72/agnew.htm.

11. Payne, *Marianne North*, 2011; Laura Ponsonby, introduction to *Abundant Beauty: The Adventurous Travels of Marianne North, Botanical Artist* by Marianne North, ed. Laura Ponsonby (Vancouver: Greystone Books, 2011).

12. Ponsonby, introduction to *Abundant Beauty*; Suzanne Le-May Sheffield, *Revealing New Worlds: Three Victorian Women Naturalists* (London: Routledge, 2001), 111–115.

13. See Payne, *Marianne North*, 14; and Lynn Merrill, *The Romance of Natural History* (Oxford: Oxford University Press, 1989),164.

14. Karen Morin, *Frontiers of Femininity: A New Historical Geography of the Nineteenth Century American West* (Syracuse, NY: Syracuse University Press, 2008), 168.

15. Morgan, *Place Matters*, 132. Morgan also describes North as a "naturalist hero" (119).

16. North, *Recollections*, 1:31.

17. Ibid., 2:99.

18. Susan Morgan reminds us that North's journal was edited and published posthumously, which means there may be descriptors in the original text—particularly comments about some of the Europeans she encountered—that were omitted by her sister before publishing. However, even given these omissions, the text is full of descriptions of plant life highlighting that plants have a more powerful influence on North than the people she meets. See Morgan's *Place Matters* for commentary about the original text and Catherine Symond's editorial work. Morgan writes: "Until someone publishes a comparison of the manuscript and the books, we cannot assess how much Catherine domesticated the memoirs by removing the obstacle of 'their very peculiar character'" (106).

19. North, *Recollections*, , 1:6–7.

20. Ibid., 84, 87.

21. Ibid., 2:28, 29.

22. Ibid., 2:137.

23. Ibid., 1:233.

24. Ibid., 2:315.

25. Ibid., 2:316.

26. Ibid., 2:26–27.

27. Ibid., 2:213.

28. Dea Birkett, *Spinsters Abroad: Victorian Lady Explorers* (London: Victor Gollancz, 1991), 21.

29. North, *Recollections*, 1:244.

30. Ibid., 157.

31. Ibid., 83.

32. Ibid., 118. Subsequent references from this volume are included parenthetically in the main text.

33. For example, she connects the Australian ranunculous plant to the bean plants she has seen in India—tracing a familial connectivity between plants and different global locations. North, *Recollections*, 2:186. Similarly, in South Africa, the aloe trees remind her of the grass trees of Australia (*Recollections*, 2:238).

34. Ibid., 2:100.

35. Patricia Murphy, *In Science's Shadow: Literary Constructions of Late Victorian Women* (Columbus: University of Missouri Press, 2006), 152.

36. Merrill, *Romance of Natural History*, 168.

37. Losano, "Preference for Vegetables," 443.

38. Marianne North, *Red Water Lily of Southern India*, painting 818 (1878), Marianne North Online Gallery, accessed February 1, 2015, http://www.kew.org/mng/gallery/818.html.

39. North, *Recollections*, 2:210-211.

40. Marianne North Online Gallery, accessed February 1, 2015, http://www.kew.org/visit-kew-gardens/explore/attractions/marianne-north-gallery.

41. North, *Recollections*, 1:48–49.

CHAPTER 4

Killer Plants of the Late Nineteenth Century

Elizabeth Chang

How might a fictional plant consume a fictional person in the nineteenth century? An unwary traveler could stumble against a trunk oozing poisonous sap while exploring a British tropical colony. A collector of exotic plants might be ensnared in his London conservatory by a choice new acquisition. And there remained always the possibility that Martian attackers could introduce carnivorous tendrils while carrying out an invasion of earth. Such attacks appeared in all kinds of Victorian fiction resistant to realism, including the popular subgenres of scientific romance, gothic horror, and colonial adventure. The obvious follow-up question—*why* might a fictional plant consume a fictional person in the nineteenth century?—has a more complicated answer. While the urgency of the plant's appetite matches the threatening taste for British flesh displayed by other monsters of the era, whether inanimate, animate, or reanimated, the idea that a plant could pursue an appetite at all defied distinctions between forms of organic life. As Henry Mayhew establishes in the final volume of *London Labour and the London Poor* (1862), "The essential quality of an animal is that it seeks its own living, whereas a vegetable has its living brought to it."[1] Fiction investigating the possibility of vegetables seeking their own living—appearing in novels, short stories, and "traveller's tales" for general audiences and avid botanists alike—therefore explored several horrors at once: first, that the expanding nineteenth-century British natural world opened up new and unexpected plant dangers, second, a plant could be *intentionally* dangerous, and third, that a plant could have any kind of intentions at all.[2]

This essay, then, is one small piece of the much larger story of nineteenth-century British literature's changing engagement with the natural world. In ways that many critics have noted, the Victorian era's realist authors, evolutionary thinkers, natural theologians, photographers, children's fabulists, botanical illustrators, sensation novelists, among many more, all found particular and signature ways to write about natural conditions both well known and newly known. Further, the new varieties of prose fiction and nonfiction that these writers and artists produced drew from each other in complex ways across generic divisions. Darwin and the novelists conversed through both form and content, explorers from Mungo Park to Francis Younghusband gave inspiration to authors of adventure tales, and, closer to home, "the eminence of the detail" linked older but persistent practices of natural description to the minutiae-laden plots of provincial novels, as Amy King has shown.[3] In many ways, then, literature of the Victorian era, from science writing to realist fiction, depended on key aspects of narrative—including characterization, setting, and descriptive and figurative language—to effect an important development. Readers of nineteenth-century fiction came to understand the natural world itself as a narrative, which could be comprehended and explained especially well using the elements of narrative.

Fringe fiction, of course, took this understanding to its extreme. Unhitched from conventions of realist representation, antimimetic fiction of the late nineteenth century used familiar narrative techniques to tell new stories of worlds much like the Victorian reader's own, with certain striking exceptions—the existence of vampires, perhaps, or the invention of time machines. These works sought to invoke a world beyond the constraints of natural laws, while always emphasizing an inevitable return to the world that those laws actually governed. Journeys to the center of the earth, to the jungles of South America, or to the depraved corners of London's East End all began with an affirmation of the scientific principles and rhetoric that the fiction would then fantastically invert and rearrange. Yet however far-ranging these narratives became, they shared a foundational premise with the more serious writing of the era described above: that the natural world could not always be comprehended in its individuality or its variety using old methods, and instead required new narrative techniques to delineate its bounds. This is especially true, I contend, when we look closely at pseudoscientific stories of carnivorous plants. While the relations between human and animal monsters has been of critical interest for some time, only recently have notions of plant sentience and intentionality directed literary or philosophical investigations.[4]

These unusual narratives are worth looking at further, however, because they compel us to ask questions about what plants can do. Stories of strange plants with anthropomorphic qualities borrowed rhetorical conventions from science writing to explain how their vegetable protagonists evolved over time and acclimatized across space. But they also boldly rewrote scientific conventions to allow a tree the carnivorous impulses of a basilisk or the higher intelligence of man himself. In this chapter, I will examine genre fiction of the late nineteenth century inspired by (though hardly faithful to) the conventions of serious scientific writing in order to follow the Victorians' developing argument for the existence of a plant-based consciousness—or more precisely, an idea of consciousness that does not explicitly exclude the possibility of plants. The challenges in such presuppositions of nonexclusion are multiple. For one, writers of fiction about conscious plants had to explain how such consciousness could be determined in the first place. For another, these fictions had also to acknowledge the variables of scale and collectivity when aligning consciousness with different kinds of plant life—fungi, forests, parasitic vines—that did not match anthropomorphic definitions of individual identity. In describing the activities of "man-eating trees," "strange orchids," and "plants that fight," this fiction used these problematic cultivars and their aggression against human characters to expand beyond the limits of realism. Plants, particularly because of their seeming resistance to fictional modes, literally ground fiction in organic experience, making their presence in realist fiction necessary to affirm diegetic bona fides. And yet to admit plants as narrative elements with any degree of agency is to defy the standard parameters by which we understand narrative fiction to operate. A plant with narrative agency radically alters notions about sentience, mobility, reproduction, and representation—not the least by blurring distinctions between character and setting.

Thus all kinds of plants that appear in fiction can work as thought experiments; evidence of environmental and aesthetic conditions that ensure that conditions within the narrative either do or do not cohere with the reader's own organic surroundings. The carnivorous plant takes this experiment further. When found in British fiction set abroad, the malevolent vegetable proposes a distributed consciousness that contravenes established relations between subject and background, but also makes it more possible to understand a hostile colonial environment as deliberately resistant and in need of broadly intrusive management or even destruction. When entering the domestic British sphere—as a commodity, treasure, stowaway, or weapon—the threatening plant's violent subversion of the imposed relationship between cultivated and cultivator not only disrupts carefully wrought alliances between humans and plants, but also suggests a disruption or revision to the notion of cultivation itself as a temporally and materially discrete process—plants may do their own cultivating, against and apart from human purpose, in a manner that both builds on and reframes better-understood linkages between female cultivation and horticultural work.[5]

Whereas late-century interest in plants has been less attended to—with Darwin's botanical writings, for example, overshadowed by his work on evolutionary theory—the Victorian discussion of plants in all their forms was vigorous, varied, and conceptually distinct from its predecessors, as this chapter will suggest. In particular, Victorian carnivorous plant fictions used popular narrative forms—the adventure novel, the mystery, the gothic horror text—to describe this new set of relationships between plants and the humans who grew them, consumed them, lived with them, and thought through them. This change was both responsive to and supportive of corresponding shifts in horticulture, botany, garden and landscape design, environmental science, natural history, and ecological consciousness taking place in the world beyond the page, which make the surrounding context for the specific changes discussed in this chapter.

Despite being only a slight piece of this larger context, man-eating plant fiction has important implications for the narrative possibilities open to a late-Victorian reading public poised on the edge of radical changes in both literary and environmental history. These changes cannot be separated from the high-water mark of imperial expansion also achieved at this time, nor from the colonial metropole's engagements with the environments of distant territories within and without the empire proper. Indeed, as multiple historians have pointed out, the expansion of the British empire was inextricably combined with environmental change, and the final years of the nineteenth century marked the strongest alignment yet between imperial expansion and ecological alteration.[6] In addition to the primary critical paradigms shaping our reading of fin de siècle literature—of the imperial gothic, of degeneration, of professionalism, of catastrophe—we can add the fictional accommodation of varieties of nonnative biological, and particularly botanical, life.[7]

Thus it is clear that plants helped late Victorian readers think about themselves and their world, in all its political, economic, and scientific expanses. Certainly Victorian empire extended itself along botanical lines in multiple ways, and in all cases, plants were increasingly understood as mobile, malleable agents of empire that enacted significant revisions in the landscape even as human activities significantly revised plant specimens and communities as well.[8] This apparently newfound plant mobility drives one interpretation of fictions about carnivorous plants: that they are allegories of the bad effects of British colonial rapaciousness in which the landscape, for once, can actually fight back against resource extraction and exploitation. However, as critics of the teleological history of plant exchanges have pointed out, environments have long been reshaped by exchanges taking place apart from the European botanical venture, so perhaps the rapacious colonizer is in this case as interested in justifying his own influence as he is in demonizing the native flora.[9]

So in addition to these stagings and restagings of the colonial encounter, we can also consider man-eating plant fictions as narrative recountings of a profound ontological inquiry. The killer plants in these stories represent threatening advancements in the ability of plants to connect with humans in a roughly equivalent way; something Robert Mitchell has termed in Romantic literature "cryptogamia" or "seduction of the human by . . . the strange and dark life of vegetation."[10] As opposed to conservationist impulses that have characterized many ecocritical readings of nineteenth-century literature, Mitchell suggests that such cryptogamia is at heart a new and transformative impulse—describing an interspecies love affair unachievable without permanent alteration on both sides.[11] Jumping ahead to the close of the nineteenth century, when Victorian botanical fictions told darker tales than Wordsworth's daffodils or Shelley's sensitive plants, we find that the alien nature of the plant continues to challenge the gap between vegetable and human through narrative.

In the remainder of this chapter, I will review the ways that plants were grudgingly awarded agency in Victorian fiction, not by virtue of their perceived suffering, but instead through fantasies of the suffering they could inflict. Scientific writing on insectivorous plants and cryptogamic fungi advanced in parallel with fictions of man-eating trees and alien red weeds, with both allowing, to greater and lesser degree, the derived intentionality of the vegetable kingdom to serve as evidence of an active and directive consciousness available somehow, though in ways not necessarily evident to humans. The search for this evidence ranges through standard colonial adventures like Frank Aubrey's *The Devil Tree of El Dorado* (1897) and later results in the most challenging and interesting fiction of this kind, produced by William Hope Hodgson, Algernon Blackwood, and H. G. Wells. Critical readings of these stories, I propose, can also shift the way we consider other genre authors of the end of the century, whether H. Rider Haggard or Arthur Conan Doyle, Robert Louis Stevenson or Bram Stoker, or even modernist forerunners Rudyard Kipling or Joseph Conrad. For all of these authors, and for countless others who also contributed works of antirealist romance, the major problem of fiction was how to perceive and understand the nature of another, particularly when that other may be self-replicating, divided, multiple, obscured, invisible, or otherwise fractured and dispersed. At the edge of modernism, fiction sought new ways to apprehend and explain the operations of an external subjectivity. Thinking about the possible thoughts of plants helped such writers work through difficulties of obscure and fractured consciousness by demonstrating just how obscure and how fractured such external subjectivities could be. Killer plant fictions depended on a concern (however sensationalized) for discernable traces of subjectivity across the foreign and colonized worlds and beyond the bounds of the human or animal body. To not consider the specific influence of the vegetable world in the rising genre of antirealist fiction is to ignore a significant piece of that genre's foundation. It is also to set aside the relevance of late-century attention to plants and plant communities as a serious concern for both art, science, and the nascent interdisciplinary field of ecology.

That the first fifty years of the nineteenth century entirely reshaped the ordinary Briton's understanding of global plant life is by now a critical commonplace. From organizations like the (later Royal) Horticultural Society, periodicals like *Gardener's Magazine* and *Gardener's Chronicle*, manuals like *The Suburban Gardener and Villa Companion* (1838), professional horticulturalists like John Lindley, and gardening experts like John and Jane Loudon and Joseph Paxton, countless amateurs learned to take seriously the study, collection, and display of plants on a broad scale.[12] Despite significant constraints of gender and class, plant appreciation and cultivation was a matter of widespread general interest and knowledge by the later nineteenth century, and authors both high- and lowbrow could deploy a range of references to plants foreign and domestic with the confidence that their readers could follow their nuances. With greater knowledge of plant names and appearances, references made to specific plants, as individual specimens, carried an epistemological heft distinguishing them from a more generally undifferentiated landscape, background, or setting. Victorian readers were taught by nonfiction and fiction alike to think seriously about their plants and the possibilities of their cultivation, in the same way that they might think about children, pets, or colonized subjects.

Of course, these are amplifications rather than inventions. Long before the publication of Charles Darwin's works *Insectivorous Plants* (1875) and *The Power of Movement in Plants* (1877), natural theology had helped readers consider the possibility that the plants they grew might have abilities or even desires unknown to their cultivator. James Tupper's *An Essay on the Probability of Sensation in Vegetables* (1811) opines that a benevolent Creator would surely "bestow upon vegetables a capacity to enjoy their own state of life,"[13] while George Towers, writing in the *Gardener's Magazine* in 1833, suggests only half-facetiously that a "plant is, *bona fide*, an organized being, endowed with sensitive life to a greater or lesser extent."[14]

Darwin's volumes, however, generated controversy for the wholly serious overlap they seemed to imply between the automatic responses of plants through actions like phototropism and the intentional responses of a conscious being. This despite Darwin's defensive claim that his general use of the term "sensitive" throughout *Insectivorous Plants* was not meant to imply consciousness but merely intended to describe the excitable glands and tentacles of the insectivorous plants he described. As he explains: "Strictly speaking, the glands ought to be called irritable, as the term sensitive generally implies consciousness; but no one supposes that the Sensitive-plant is conscious, and as I have found the term convenient, I shall use it without scruple."[15] As historians of science have pointed out, Darwin's language links him to earlier, discredited experimentation on photo- and gravitropism and sensitivity in plants that posited that such reactions established a chain of association between plant and animal behavior and vitalism.[16] Even more troubling to fellow scientists were the assertions of Darwin's German follower Ernst Haeckel, who, under the aegis of scientific monism, concluded that since the movements of the sensitive plants are "strikingly similar to the movements of the lower animal forms: whoever ascribes consciousness to the latter cannot refuse it to such vegetal forms."[17]

For indeed, despite the scruples of established science, many were ready to suppose that a plant could be conscious and that consciousness and even morality could be implied from plant movement. H. Rider Haggard, a prolific and enthusiastic gardener and agricultural writer when not chronicling the adventures of Allan Quartermain (*A Farmer's Year* [1899], *Rural England* [1902], *A Gardener's Year* [1905]), describes an unusual Sundew specimen in his collection as a "vegetable butcher," and continues:

> To my mind, its unpleasant habits show in a very striking manner how real, if subtle, is the connection between the animal and the vegetable world, for here we have a plant actually feeding on the living creatures that it has caught, and, what is more, baiting its traps in order to catch them. Is there, then, so wide a gulf between it and *homo sapiens*, who does precisely the same thing and lives thereby? We think nothing of putting this law of death—Nature's hideous scheme—in motion for our own profit, but when a wretched little plant imitates our exalted example, the effect is uncanny.[18]

The step forward made here from plant sensitivity to plant malevolence was long anticipated, as Theresa Kelly has shown in her discussion of Erasmus Darwin's consideration of the venomous Upas tree.[19] Haggard's move to the propositional sphere ("Is there, then . . .") may seem particularly apt for a writer of fantastic fiction, but he was certainly not the only writer to wonder if the strong distinction between plant "habits" and human actions might in fact be only a perceptual construct.[20]

The carnivorous sundew also attracted members of the aesthetic and decadent movement, a group predisposed to favor unsettling interchanges between plant and animal life.[21] Algernon Swinburne's poem "The Sundew" (1862) similarly reflects on the ontological proximity between humans and plants: "You call it sundew; how it grows / If with its colour it have breath, / If life taste sweet to it, if death / Pain its soft petal, no man knows: / Man has no sight or sense that saith" (ll. 26–30).[22] Later, Grant Allen, a science writer of some influence before his blossoming as an author of detective and New Woman fiction, drew on both Darwin and Swinburne in calling the plant "atrociously and deliberately wicked" in his 1884 article "Queer Flowers," written for the *Cornhill Magazine*.[23] Allen frames the sundew's description in a larger reflection on the horrors of possible plant sentience, writing, "There is something too awful and appalling in this contest of the unconscious and insentient with the living and feeling, of a lower vegetative form of life with a higher animated form," continuing, "There seems to be a sort of fiendish impersonal cruelty about its action which sadly militates against all our pretty platitudes about the beauty and perfection of living beings."[24] Allen's distress over the sundew inverts customary directions of sympathy by doubly deploying "murderous propensities."[25] The sundew itself is actively cruel in seeking out its insect prey, but the sundew's cruelty is also metaphorically active, "militating" its way into conventional discourse and disrupting the familiar systems of figurative language. In both regards, the sundew, an "inconspicuous small weed" with "literary and scientific honours . . . heaped upon its head to an extent almost unknown in the case of any other member of the British floral commonweal" (404), goes against the standards by which fin de siècle Victorian culture has been held to recognize conscious existence: the sundew impresses not because it suffers itself, but because it causes other to do so.[26] Counterpoised to late-century debates over animal welfare, vivisection, and vegetarianism, discussions of plant sentience now gravitated not to sympathy but to antipathy and fear.[27]

Even in Samuel Butler's satire of colonial adventure fiction *Erewhon* (1872), plant consciousness is memorably detailed through the "low cunning" of a potato in a dark cellar. Appearing within its intertext "Book of the Machines," the description of the potato's advances accommodates both a mockery of the Erewhonian professor of botany's squeamish attention to vegetable rights and a surprisingly activist assertion of plant volition. Of the cunning potato, Butler writes: "He [the potato] knows perfectly well what he wants and how to get it. He sees the light coming from the cellar window and sends his shoots crawling straight thereto . . . we can imagine him saying, 'I will have a tuber here and a tuber there, and I will suck whatsoever advantage I can from all my surroundings.' . . . The potato says these things by doing them, which is the best of languages. What is consciousness if this is not consciousness?"[28] Though Butler uses this example largely as an opportunity for comedy at the expense of the Erewhonians, such satire does not negate his equally radical expansion of consciousness's proof-case. Nor, despite the obviously false imposition of a personal pronoun and personifying form, does Butler intend us to not accept the consequences for agency that the potato's advancing tubers propose, for such ideas resonate across his work. Philip Armstrong has pointed out that the language of the Erewhonian professor of botany is borrowed in part from Butler's own lecture "The Subdivision of the Organic World into Animal and Vegetable," and

Gillian Beer has shown that Butler's "pleasure in imagining the eagerness of other life forms" is reflected in his nonfiction studies.[29] *Life and Habit* (1878), for example, proposes that personal identity in humans rests upon the same self-generating volition that allows "the lichen . . . [to] grow upon the granite rock by first saying to itself, 'I think I can do it.'"[30] As Butler investigates, personhood is equally a continuously evolving condition linking one life stage with the next and at the same time a singular and momentary expression of consciousness; in much the same way, even as carnivorous plant stories showed the consequences of natural selection advancing at terrible speed toward a horrifying end, they also illuminated an asynchronous notion of plant intentionality apprehended only in the moment.

This was as true when plant predators were used as a blunt instrument of horror as much as when they were held to offer a nuanced exploration of vegetable consciousness. Low-quality adventure stories, aimed at readers unaccustomed to considering Allen's and Butler's philosophical and moral concerns, framed the problem of discerning plant intentions as a matter of self-preservation amid an uncontrolled and fecund environment. The young Arthur Conan Doyle's 1879 story "An American's Tale," Phil Robinson's "The Man-Eating Tree" (1881), Frank Aubrey's *The Devil Tree of El Dorado* (1897), and Fred White's "Purple Terror" (1899) all present vicious trees (or, in Doyle's case, tree-sized murderous flytraps) in locations both lush and indeterminate. Doyle's flytraps grow in a frontier alternately identified as Arizona and Montana where "Grass as hung over a chap's head as he rode through it, and trees so thick that you couldn't catch a glimpse of blue sky for leagues and leagues, and orchids like umbrellas!"[31] Roraima, the great plateau that conceals both Aubrey's devil tree and the long-lived tribe that keeps the tree's secrets, contains "flora and fauna [that] flourish unchecked in the utmost luxuriance of tropical savage life," where, bafflingly enough to the British narrator, "one of the greatest marvels and mysteries of the earth lies on the outskirt of one of our colonies, and we leave the mystery unsolved, the marvel uncared for."[32] Will Scarlett, the enterprising amateur botanist and central character of White's story, first experiences the "Purple Terror" in a military expedition across Cuba, where his "geographical and botanical knowledge were going to prove of considerable service to a grateful country when said grateful country should have passed beyond the rudimentary stages of colonization."[33] Cheryl Blake Price has shown in her work on man-eating trees that these stories reflect both "ecophobic reactions to the colonial environment" and "anxiety that the colonial wilderness . . . was disappearing through the mechanisms of colonialism," and there are many further implications for the consideration of animate organic landscape in these fictions that go beyond my particular focus on plant consciousness.[34]

So even as these landscapes are established as visually and horticulturally unintelligible or unknown, their organic elements are also constantly evolving into new forms. Robinson's narrator, recounting a brush with a man-eating tree in central Africa, begins not with the exciting plot details of the attack, but with an extended philosophical reflection on the necessary correspondence between plant and animal worlds, concluding:

> The vegetable world, however, has its revenges. You may keep the guinea pig in a hutch, but how will you pet the basilisk? The little sensitive plant in your garden amuses your children . . . but how could you transplant a vegetable that seizes the running deer, strikes down the passing bird, and once taking hold of him, sucks the carcass of man himself, till his matter becomes as vague as his mind, and all his animate capabilities cannot escape him from the terrible embrace of—God help him!—an inanimate tree?[35]

Kelly Hurley, exploring fin de siècle gothic, has proposed that Aubrey, Doyle, and Robinson are of a piece with William Hope Hodgson and H. G. Wells in imagining evolutionary monstrosities, animal and vegetable, that generally attack the priority of the specific human form: "The viscosity of the predatory natural world may be said to represent the suchness of matter, as it gains sentience and rises up to swallow the bounded human world," she suggests.[36] Hurley's reading is borne out by Robinson's insistence that "the sensual instincts of beast and vegetable are manifestly analogous—the world must be as percipient as sentient throughout" and, equally, by his claim that "given the necessity of . . . urgent self-interest, every animal or vegetable could eventually revolutionize its nature."[37] But Robinson's identification of the particularly terrible revenges of the vegetable world links his evolutionary fantasy not just to any kind of monster but to monsters emerging from practices of Victorian plant enthusiasms' global plant exchange specifically. The fear is not that the petted plant specimens in the kitchen garden will evolve *into* sentience, but that they possess sentience already.

Equally, the desires of Will Scarlett, who is lured to the "Purple Terror" both by his lust for new plants and by his lust for a beautiful woman, make him vulnerable to attack in ways unique to the horticulturally enraptured Victorian era. When he viewed the flowers of the murderous parasitical vine, "All Scarlett's scientific enthusiasm was aroused. It is not given to every man to present a new orchid to the horticultural world. And this one would dwarf the finest plant hitherto discovered."[38] While White's story only proposes such an exchange—the orchids remain rooted in their Caribbean setting—it was also a common feature in tales of murderous plants used to describe the effects of a specimen returned to a domestic locale. These acts of botanical exchange were not unique to the era, but the vast increase in the scale of the collections and the reach of the botanical collectors placed ever greater pressure on narrative and nomenclature to preserve geographical distinction between native and nonnative species. As John Rieder points out, such "fantasies of appropriation," cloaked as "zoological and ethnological acquisitiveness," unite the emerging genre of science fiction with the earlier prose of travel narrative under the logic of colonialism; given the overwhelming evidence of Victorian emotional attachment to their plants, we must also add to these appropriations the fantasy of botanical acquisition.[39]

As Lynn Voskuil shows elsewhere in this volume in her reading of H. G. Wells's "The Flowering of the Strange Orchid" (1904), such acquisitiveness left domestic collectors ripe for dissection by both the horror storyteller and the satirist. Another, less well-studied direction for modernist horror examines what happens to questions of vegetable sentience and malevolence when the antagonist is not an easily anthropomorphized tree, but instead a diffuse collection of fungal spores. Examples of late nineteenth-century fungus horrors include John Uri Lloyd's *Etidorhpa* (1895) and William Hope Hodgson's influential 1907 short story "The Voice in the Night," first published in the popular and influential pulp magazine *Blue Book.* Hodgson, a former member of the merchant marine whose works of horror were frequently set a sea, tells the story of an ill-fated pair of lovers who encounter an island where a "vile fungus . . . was growing riot. In places it rose into horrible, fantastic mounds, which seemed almost to quiver, as with a quiet life, when the wind blew across them. Here and there, it took on the forms of vast fingers, and in others it just spread out flat and smooth and treacherous . . . [t]he whole quaking vilely at times."[40] The gradual realization that the fungus is growing unstoppably, not just across the island but through and around their bodies, is followed by a sudden and insurmountable compulsion to eat the growth. This desire continues even after encountering "an extraordinarily shaped mass of fungus . . . swaying uneasily, as though it possessed life of its own" with "a grotesque resemblance to the figure of a distorted human creature," which, upon grim consideration, is understood to in fact be a sailor previously marooned on this same island. In understanding and nobly accepting their future fate, the couple's narration spares the tale's shipboard listeners, who already perceive the fiancé of the pair as no more than a "great, gray nodding sponge."[41] The story's horror, then, comes not from the fear of dying on the fungus island, but of continuing to live there—albeit in a greatly transformed fashion. Fantasies of unification between plant and human continue to chill precisely because they place the resulting hybrid at the far outer limits of not only narrative, but consciousness itself.

Hodgson returned to the genre of the malevolent sea-plant in his later short story "The Derelict," but it was "The Voice in the Night" that proved most influential to later fungal fictions, including Philip M. Fisher's "Fungus Isle" (1923). Fungi, and in particular cryptogamic fungi, had already been a subject of interest and repulsion throughout the second half of the century, especially because their spore-based reproduction resisted so strongly conventional structures of metaphor.[42] It seemed impossible to distinguish parasitic fungi from their hosts, let alone identify the singular personhood of the fungus itself. Yet larger plant forms could also form a distributed consciousness with expansive and uncertain bounds. One of the most effective of all murderous plant fictions, Algernon Blackwood's 1907 short story "The Willows," describes the near-sacrifice of two travelers on the Danube to an island of psychically manipulative willows. In the narrator's horrified realization of the evil at work, plants are made mobile and humans fixed in place: "Creeping with silent feet over the shifting sands, drawing imperceptibly nearer by soft unhurried movements, the willows had come closer during the night. . . . There was a suggestion here of personal agency, of deliberate intention, of aggressive hostility, and it terrified me into a sort of rigidity."[43] The psychological shock of managing a marooning in a place where the characters face hostile vegetation—"We touched the frontier of a region where our presence was resented. . . . We were the first human influences on this island and we were not wanted. *The willows were against us*"—paralyzes the narrator into a helpless inability to save himself from his fate.[44] Though the two protagonists do eventually escape, spared by the sacrificial murder (apparently by willow) of an anonymous peasant, the effect of an environment apparently "on the frontier of another world, an alien world, a world tenanted by willows only and the souls of willows," shows at what cost the human world is defended against a botanical villainy far more ontologically complex than Aubrey's devil tree.[45] "The Willows," one of the earliest and most widely read of Blackwood's stories, clearly illustrates his lifelong interest in spiritual, mystical and psychic extremes, but also obviously depends upon his finely observed notions of the elements of landscape and plants in particular to achieve its effective horrors. There is no longer any question that plants possess both a metaphorical and actual agency; inquiry is instead redirected to the inadequacy of metaphor to defend the individual human psyche against the incursions of the natural world.

If Blackwood's macabre modernism still stands as the best example of plant malevolence in the twentieth century, H. G. Wells's *The War of the Worlds* remains remarkable for its nineteenth-century naturalization of the most alien of vegetable attackers. Wells does not foreground the dual nature (animal and vegetable) of the Martian invaders in his novel, and in fact we do not learn of the "red weed," the plant that gives Mars its characteristic color and threatens to choke England in the process of doing the same, until we are far into Book 2, "The Earth under the Martians." Even then the narrator's introduction is presented as an allusion within a more general discussion of the differences between life on Mars and terrestrial life. He records:

> At any rate, the seeds which the Martians (intentionally or accidentally) brought with them gave rise in all cases to red-coloured growths. Only that known popularly as the Red Weed, however, gained any footing in competition with terrestrial forms. The Red Creeper was quite a transitory growth, and few people have seen it growing. For a time, however, the Red Weed grew with astonishing vigour and luxuriance. It spread up the sides of the pit by the third or fourth day of our imprisonment, and its cactus-like branches formed a carmine fringe to the edges of triangular window. And afterwards I found it broadcast throughout the country, and especially wherever there was a stream of water present.[46]

This passage is notable especially for its dislocations of scale between, on the one hand, the broad horticulturally omniscient description given of the red weed's "vigor and luxuriance," elsewhere called "titanic" and "gigantic and of unparalleled fecundity," and, on the other, the first-person narrator's limited knowledge about the weed, which initially matches exactly his limited visual perspective, trapped in a bombed house with a terrified curate and with only that triangular window framed in red weed to observe the horrifying progress of the Martian invaders.[47]

Similarly, he tethers his account of the weed to the deflating temporal reversals of the passage, with the offhand use of the word "afterwards" spoiling prematurely the inevitability of the invasion's failure. In introducing the weed at this point, the narrator has already cut away from a vivid description of the arrival of the fifth Martian cylinder to "add in this place certain further details which, although they were not all evident to us at the time, will enable the reader who is unacquainted with them to form a clearer picture of these offensive creatures."[48] The narrator's reluctance to comply with the rules of his own narrative condition and tell the story as it occurred here corresponds with novel's far more mysterious model of narrative agency: that of the Martians themselves, both the creatures operating the tripods and the swiftly growing fronds of the red weed.

As the narrator repeatedly reminds us, the British subjects involved never really know why the Martians come to Earth but must instead endlessly speculate about what the Martians' known actions say about their probable intent. But despite the lack of human understanding of Martian volition, there are multiple intentional actors here. The tripod operators are of course viciously active, but we also see that red weed is itself exercising a form of mobility that appears to be intentional. This apparent intentionality on the part of this invasive plant is, in fact, of critical importance given the most central concerns of the novel. In *War of the Worlds,* this weaker sort of derived intentionality is the only one admissible either when thinking about narrative agency or when inferring the presence of a threatening and otherwise inaccessible alien consciousness. The British waterways that carry the red weed and the mechanical tripods that transport the Martian creatures are both prostheses for the alien invaders, amplifying and making more legible the movements that assert directive consciousness.

Though the creatures in tripods accept more readily than the red weed the impositions of anthropomorphism, Wells does not functionally distinguish between the two. This supports the idea that Wells implicitly advances throughout his work—that plants in general, and this plant in particular, represent an outer limit to the range of human interest in the alien.[49] Indeed, the distinction between the two kinds of Martians is largely irrelevant: the young Martians "bud off" their parents, like "young lilybulbs" or "young animals in the fresh-water polyp"; the older Martians, lacking entrails (or any organs besides brains and hands) are, plantlike, sustained by fresh blood obtained "directly by means of a little pipette into the recipient canal," and, in short, act very much like a superintelligent version of the red weed itself.[50] That Wells is, throughout his fiction, concerned with the distinctions between jungles and gardens is of a piece with his particular interest in weeds—the alien of plants, and the best example of the alien nature of plants. This is, of course, because "weed" is exclusively a derived and relative term—it carries meaning only in relation to some other class of things that are not weeds—and so to recognize a weed is to recognize an arbitrary distinction of purpose that the weed by its lively vigor aims to resist.

That the novel uses the Martians to critique, via reverse colonization, the follies of British imperialism is, of course, a touchstone of its reception and its era, as Stephen Arata and others have shown.[51] And clearly the novel can and should be read, in part, as an account of a galactic version of an acclimatization society that has gone terribly wrong.[52] It also serves as an example of global circulation that divorces the native and natural and deprives the horticultural specimen of its proper local environment. As this novel seeks to demonstrate, even plants that do not survive by attacking unwitting travelers have intentions that the observer cannot register except by effect, and even then only if that effect happens to be that of causing grievous harm. Human hands—implied constants in the actions of broadcasting, acclimatizing, and transplanting—falsely imply human agency in the development of global environments, but Wells means to pick apart the mental constructs that support that false narrative of exclusively human influence on the surrounding world. At the dawn of what we now call the environmental movement, this novel offers the idea of the functionally sentient plant as an example of the way in which environments and plants can, in fact, broadcast themselves, through a directive consciousness that can both be understood to exist and yet be defiantly and totally inaccessible to our figures and narrative forms. When Wells's narrator recounts a walk through the ruined landscape as "all about me the red weed clambered among the ruins, writhing to get above me in the dimness," his insistently foregrounded imposition of intent onto the plant hints at the countless other impositions and violent figurative replacements that British fictions have done to their fictional vegetable worlds.[53]

Adela Pinch has recently pointed out, in her study of the priority of other people's thoughts to readers and writers of the nineteenth century, that "we have grown accustomed to thinking of Victorian Britain as a realm of science, but it was also a realm of metaphysical speculation."[54] When thinking about the thoughts of plants, Victorian writers and readers found a double redirection: in one sense, plants grew narratively more vigorous as they gained agency and direction commensurate with animals and even monstrous or villainous humans, while in another sense, plants grew inaccessible to narrative and rhetorical figures as they became weighted with specific and situated forms of geographical and scientific knowledge. Environmental dramas restaging the natural world, like these strange science fictions of malevolent plants, show just some of the many paths to modernism's challenge of the clean narrative distinction between self and other. But they also suggest ecological fiction's persistent challenge: to imagine how the organic world imagines the human is also to confront the limits of the possibilities of imagination.

Notes

1. Henry Mayhew, *London Labour and the London Poor*, vol. 4, *Those That Will Not Work* (London: Charles Griffin, 1862), 2.

2. Recent literary studies of killer plants include Cheryl Blake Price's "Vegetable Monsters: Man-Eating Trees in *Fin-de-Siècle* Fiction," *Victorian Literature and Culture* 41, no. 2 (2013): 311–27, and T. S. Miller, "Lives of the Monster Plants: The Revenge of the Vegetable in the Age of Animal Studies," *Journal of the Fantastic in the Arts* 23, no. 3 (2012): 460–79. Also very useful to this study have been Price's Internet archive of man-eating plant fictions, "Bibliographic Resources on 'Man Eaters' in Victorian Literature," *Cheryl Blake Price Academic Website*, accessed November 15, 2011, http://www.cherylblakeprice.com/my-research/bibliographic-resources-on-man-eaters-in-victorian-literature/), and the popular collections edited by Chad Arment, *Flora Curiosa: Cryptobotany, Mysterious Fungi, Sentient Trees, and Deadly Plants in Classic Science Fiction and Fantasy* (Landisville, PA: Coachwhip Publications, 2008) and *Botanica Delira: More Stories of Strange, Undiscovered, and Murderous Vegetation* (Landisville, Pa: Coachwhip Publications, 2010). See also two indexes of early science fiction: Everett Franklin Bleiler and Richard Bleiler, *Science-Fiction, the Early Years: A Full Description of More Than 3,000 Science-Fiction Stories from Earliest Times to the Appearance of the Genre Magazines in 1930* (Kent, OH: Kent State University Press, 1990) and Brian M. Stableford's *Science Fact and Science Fiction: An Encyclopedia* (New York: Taylor & Francis, 2006).

3. Amy M. King, "Reorienting the Scientific Frontier: Victorian Tide Pools and Literary Realism," *Victorian Studies* 47, no. 2 (2005): 158. On exchanges with Darwin, see Gillian Beer, *Darwin's Plots: Evolutionary Narrative in Darwin, George Eliot, and Nineteenth-Century Fiction* (Cambridge: Cambridge University Press, 2000); George Levine, *Darwin and the Novelists: Patterns of Science in Victorian Fiction* (Cambridge: Harvard University Press, 1988); and Gowan Dawson, *Darwin, Literature, and Victorian Respectability* (Cambridge: Cambridge University Press, 2007). On the intersections of travel literature and scientific exploration, see Tim Fulford, Debbie Lee, and Peter Kitson, *Literature, Science and Exploration in the Romantic Era: Bodies of Knowledge* (Cambridge: Cambridge University Press, 2004); Paul Smethurst, *Travel Writing and the Natural World, 1768–1840* (New York: Palgrave Macmillan, 2012); Peter Raby, *Bright Paradise: Victorian Scientific Travellers* (Princeton, NJ: Princeton University Press, 1997); and Mary Louise Pratt, *Imperial Eyes: Travel Writing and Transculturation* (New York: Routledge, 2007).

4. Among philosophers of botany, see Sylvie Pouteau, "Beyond 'Second Animals': Making Sense of Plant Ethics," *Journal of Agricultural and Environmental Ethics* 27, no. 1 (2014): 1–25; Val Plumwood, *Environmental Culture: The Ecological Crisis of Reason* (New York: Routledge, 2001); Matthew Hall, *Plants as Persons: A Philosophical Botany* (Buffalo: SUNY Press, 2011); and John Charles Ryan, "Passive Flora? Reconsidering Nature's Agency through Human-Plant Studies (HPS)," *Societies* 2, no. 3 (2012): 101–21.

5. On these, see Amy M. King, *Bloom: The Botanical Vernacular in the English Novel* (Oxford: Oxford University Press, 2003); and Deidre Shauna Lynch, "'Young Ladies Are Delicate Plants': Jane Austen and Greenhouse Romanticism," *ELH* 77, no. 3 (2010): 689–729.

6. See William Beinart's companion to the *Oxford History of the British Empire: Environment and Empire* (Oxford: Oxford University Press, 2007) as well as the following key histories: Richard Grove, *Green Imperialism: Colonial Expansion, Tropical Island Edens, and the Origins of Environmentalism, 1600–1860* (Cambridge: Cambridge University Press, 1995); Richard Harry Drayton, *Nature's Government: Science, Imperial Britain, and the "Improvement" of the World* (New Haven: Yale University Press, 2000); Lucile Brockway, *Science and Colonial Expansion: The Role of the British Royal Botanic Gardens* (New York: Academic Press, 1979); Ray Desmond, *The History of the Royal Botanic Gardens Kew* (Kew: Kew Publications / Royal Botanic Gardens, Kew, 2007); John M. MacKenzie, ed., *Imperialism and the Natural World* (Manchester: Manchester University Press, 1990); Donal P. McCracken, *Gardens of Empire: Botanical Institutions of the Victorian British Empire* (London: Leicester University Press, 1997); and David Arnold, *The Tropics and the Traveling Gaze: India, Landscape, and Science, 1800–1856* (Seattle: University of Washington Press, 2006).

7. For these paradigms, see Patrick Brantlinger, *Rule of Darkness: British Literature and Imperialism, 1830–1914* (Ithaca, NY: Cornell University Press, 1988); Stephen Arata, *Fictions of Loss in the Victorian Fin de Siècle* (Cambridge: Cambridge University Press, 2009); Nicholas Daly, *Modernism, Romance and the Fin de Siècle: Popular Fiction and British Culture, 1880–1914* (Cambridge: Cambridge University Press, 1999); and Cara Murray, "Catastrophe and Development in the Adventure Romance," *English Literature in Transition, 1880–1920* 53, no. 2 (2010): 150–69.

8. For more on such transfers, see Eric Pawson, "Plants, Mobilities and Landscapes: Environmental Histories of Botanical Exchange," *Geography Compass* 2, no. 5 (2008): 1464–77; Jill H. Casid, *Sowing Empire: Landscape and Colonization* (Minneapolis: University of Minnesota Press, 2005); and William Beinart and Karen Middleton, "Plant Transfers in Historical Perspective: A Review Article," *Environment and History* 10, no. 1 (2004): 3–29.

9. As Rangan, Carney, and Denham have argued, both narratives of "ecological imperialism," focusing on the destruction wreaked on both colonizing and colonized environments by introduced species, and narratives of "ecological nationalism," concentrating on the economic development of modern states enabled by plant exchange, equally depend on an "inherent Eurocentrism" that "tends to obscure the fact that plant transfers have been an integral part of human history, extending over several millennia through the quotidian interactions of provisioning and exchange between peoples in every part of the world." See Haripriya Rangan, Judith Carney, and Tim Denham, "Environmental History of Botanical Exchanges in the Indian Ocean World," *Environment and History* 18, no. 3 (2012): 312.

10. Robert Mitchell, "Cryptogamia," *European Romantic Review* 21, no. 5 (2010): 632.

11. Here Mitchell is in conversation with such field-defining works as Jonathan Bate's *Romantic Ecology: Wordsworth and the Environmental Tradition* (New York: Routledge, 1991) and other works of so-called green romanticism that advocate not for transformation but for preservation.

12. On these, see Ann B. Shteir, *Cultivating Women, Cultivating Science: Flora's Daughters and Botany in England, 1760–1860* (Baltimore: Johns Hopkins University Press, 1996) and writings on amateur naturalist societies including Anne Secord, "Corresponding Interests: Artisans and Gentlemen in Nineteenth-Century Natural History," *British Journal for the History of Science* 27, no. 4 (1994): 383–408.

13. James Perchard Tupper, *An Essay on the Probability of Sensation in Vegetables: With Additional Observations on Instinct, Sensation, Irritability, &c* (London: White, Cochrane, 1811), 80–81.

14. G T [George Towers], "Fertility Induced by Spur-Pruning," *Gardener's Magazine* 9 (1833): 319.

15. Charles Darwin, *Insectivorous Plants* (London: John Murray, 1875), 19.

16. On this history, see Craig W. Whippo and Roger P. Hangarter, "The 'Sensational' Power of Movement in Plants: A Darwinian System for Studying the Evolution of Behavior," *American Journal of Botany* 96, no. 12 (2009): 2115–27; as well as Soraya De Chadarevian, "Laboratory Science versus Country-House Experiments: The Controversy between Julius Sachs and Charles Darwin," *British Journal for the History of Science* 29, no. 1 (1996): 17–41. Chadarevian points out that the review of Darwin's *The Power of Movement in Plants* in the *Times* suggests broad popular interest in Darwin's long-standing botanical inquiries accompanying his fame as an evolutionary theorist (21).

17. Ernst Heinrich Philipp August Haeckel, *The Riddle of the Universe at the Close of the Nineteenth Century* (New York: Harper & Brothers, 1905), 177.

18. Henry Rider Haggard, *A Gardener's Year* (London: Longmans, Green, 1905), 243.

19. Theresa M. Kelley, *Clandestine Marriage: Botany and Romantic Culture* (Baltimore: Johns Hopkins University Press, 2012), 86–87.

20. It has been pointed out that a full definition of the conditions of botanical carnivory has never been achieved; see Mark W. Chase et al., "Murderous Plants: Victorian Gothic, Darwin and Modern Insights into Vegetable Carnivory," *Botanical Journal of the Linnaean Society* 161, no. 4 (2009): 329–56.

21. On this, see Barbara Larson, "Evolution and Victorian Art," in *Evolution and Victorian Culture*, ed. Bernard Lightman and Bennett Zon (Cambridge: Cambridge University Press, 2014), 121–48.

22. For more on Swinburne's poem and its relations to Darwin's text, see Jonathan Smith, "Une Fleur du Mal? Swinburne's 'The Sundew' and Darwin's 'Insectivorous Plants,'" *Victorian Poetry* 41, no. 1 (2003): 131–50.

23. Grant Allen, "Queer Flowers," *Cornhill Magazine* 3, no. 16 (1884): 404. Just prior to the appearance of this article, Allen had published two noted botanical studies: *The Colours of Flowers* (London: Macmillan, 1882) and *Flowers and Their Pedigrees* (London: Longmans, 1883).

24. Allen, "Queer Flowers," 403, 404.

25. Ibid., 404.

26. Ibid.

27. On these debates, see Jed Mayer, "The Expression of the Emotions in Man and Laboratory Animals," *Victorian Studies* 50, no. 3 (2008): 399–417, among many others.

28. Samuel Butler, *Erewhon; or, Over the Range*, ed. Hans-Peter Breuer and Daniel F. Howard (Newark: University of Delaware Press; London: Associated University Presses), 184.

29. Gillian Beer, "Butler, Memory and the Future," in *Samuel Butler, Victorian against the Grain: A Critical Overview*, ed. James G. Paradis (Toronto: University of Toronto Press, 2007), 47; Philip Armstrong, "Samuel Butler's Sheep," *Journal of Victorian Culture* 17, no. 4 (2012): 450.

30. Samuel Butler, *Life and Habit* (London: Trübner, 1878), 82. No modern edition of this text exists; the 1981 Wildwood House edition is a reprint of the 1910 text.

31. Sir Arthur Conan Doyle, "The American's Tale," in *The Best Science Fiction of Arthur Conan Doyle*, ed. Charles G. Waugh and Martin H. Greenberg (Carbondale: Southern Illinois University Press, 1981), 4.

32. Frank Aubrey, *The Devil-Tree of El Dorado: A Novel* (New York: New Amsterdam Book Co., 1897), 273. For more on Roraima as a site for adventure narrative, see also R. G. Forman, "Room for Romance: Playing with Adventure in Arthur Conan Doyle's *The Lost World*," *Genre* 43, nos. 1–2 (2010): 27–59.

33. Fred White, "The Purple Terror," *Strand Magazine* 18, no. 105 (1899): 243.

34. Price, "Vegetable Monsters," 327.

35. Phil Robinson, "The Man-Eating Tree," in *Under the Sun* (Boston: Roberts Brothers, 1882), 299.

36. Kelly Hurley, *The Gothic Body: Sexuality, Materialism, and Degeneration at the Fin de Siècle* (Cambridge: Cambridge University Press, 1996), 62.

37. Robinson, "The Man-Eating Tree," 297.

38. White, "The Purple Terror," 245.

39. John Rieder, *Colonialism and the Emergence of Science Fiction* (Middletown, CT: Wesleyan University Press, 2008), 56.

40. William Hope Hodgson, "A Voice in the Night," in *Men of Deep Waters* (London: Holden & Hardingham, 1921), 178.

41. Ibid., 182, 184.

42. On one female botanist and social reformer's debates with Darwin over such fungi, see Tina Gianquitto, "Botanical Smuts and Hermaphrodites: Lydia Becker, Darwin's Botany, and Education Reform," *Isis* 104, no. 2 (2013): 250–77.

43. Algernon Blackwood, "The Willows," in *The Listener and Other Stories* (London: Eveleigh Nash, 1907), 160.

44. Ibid., 147.

45. Ibid., 145.

46. H. G. Wells, *A Critical Edition of the War of the Worlds: H. G. Wells' Scientific Romance*, introduction and notes by David Y. Hughes and Harry M. Geduld, (Bloomington, IN: Indiana University Press, 1993), 151–52.

47. Ibid., 166.

48. Ibid., 150.

49. On this see Emily Alder, "'Buildings of the New Age': Dwellings and the Natural Environment in the Futuristic Fiction of H. G. Wells and William Hope Hodgson," in *H. G. Wells: Interdisciplinary Essays*, ed. Steven McLean (Aldershot: Cambridge Scholars, 2008), 114–129.

50. Wells, *War of the Worlds*, 150.

51. Arata notes, "H. G. Wells located the germ of his *War of the Worlds* in a discussion with his brother Frank over the extermination of the indigenous Tasmanian population under British rule," in "The Occidental Tourist: 'Dracula' and the Anxiety of Reverse Colonization," *Victorian Studies* 33, no. 4 (1990): 623.

52. For more on the red weed as invasive plant species, see Christina Alt, "Prickly Pears and Martian Weeds: Ecological Invasion Narratives in History and Fiction," in *Rethinking Invasion Ecologies from the Environmental Humanities*, ed. Jodi Frawley and Iain McCalman (New York: Routledge, 2014), 137–48. See also Matthew Chew's brief summary of the late-nineteenth-century meteoric rise and sudden collapse of the canal-choking Canada waterweed (*Elodea*) in Britain; he concludes, "What can be made of such capriciousness? Fiction. *Elodea* doubtless inspired the 'red weed' that accompanied H. G. Wells' Martians to England in *The War of the Worlds*," "The Monstering of Tamarisk: How Scientists Made a Plant into a Problem," *Journal of the History of Biology* 42, no. 2 (2009): 235.

53. Wells, *War of the Worlds*, 183.

54. Adela Pinch, *Thinking about Other People in Nineteenth-Century British Writing* (Cambridge: Cambridge University Press, 2010), 6–7.

PART II

Strange Bodies

Rethinking Physiology

CHAPTER 5

Reading through Deafness

Francis Galton and the Strange Science of Psychophysics

Danielle Coriale

One of the strangest Victorian sciences originated in the work of Gustav Theodor Fechner, a German experimental psychologist who believed that the relationship between "body and soul" could be studied with mathematical precision.[1] Rather than accept that an insuperable barrier distinguishes that which is "mental, psychological, or belonging to the soul" from that which is "bodily, corporeal, physical, or material," Fechner sought to study human beings from a "single point of view."[2] In his monumental book *Elemente der Psychophysik* (1860), he argued that scientists could move beyond the mind-body distinction by studying sensory experience, which involves the physical stimulation of nerves and the mental recognition of the feeling produced by that stimulation. Having coined the term *Psychophysik* (translated into English as *psychophysics*) to describe this new line of scientific inquiry, Fechner designed rigorous experiments that would allow him to exploit its full potential.[3] In these experiments, which he describes at length in *Elemente,* he increased the intensity of a stimulus in small increments and then recorded the subject's perception of those increases in numerical terms. By comparing the two figures, Fechner determined the mathematical relationship—

which he referred to as the "psycho-physical law"—between physical stimulation and psychological experience of a sensation.[4]

Fechner's development of psychophysics had dramatic and enduring effects on the field of psychology in the late nineteenth century. Soon after *Elemente der Psychophysik* appeared in 1860, German psychologists Wilhelm Wundt and Hermann Ebbinghaus incorporated Fechner's theories and psychophysical experiments into their research on sensation. And although *Elemente* was not translated into English until 1966, the book made a powerful impression on scientists around the world as reviews and excerpts circulated in British, French, and American periodicals during the early 1870s. By 1875, Francis Galton praised Fechner's book for "lay[ing] the foundations of a new science" and noted that psychophysics was "beginning to attract serious attention in Belgium, France, America and England."[5] By the time William James published his *Principles of Psychology* in 1890, *Elemente der Psychophysik* had made its mark on the field of psychology. As James observed, "Fechner's book was the starting point of a new department of literature, which it would be perhaps impossible to match for the qualities of thoroughness and subtlety."[6] Although James went on to criticize Fechner's "peculiarly fragile" methods in *Principles*, psychophysics did not linger long on the periphery of mainstream science.[7] During the early twentieth century, Fechner's once-strange science gained a stronghold in psychology and remains an important field in the discipline today.[8] Finally, Fechner's innovative way of thinking about the mind-body problem inspired twentieth-century scientists and philosophers including Sigmund Freud, Henri Bergson, Alfred Whitehead, Ernst Mach, and Ilya Prigogine—all of whom acknowledged his profound influence on their work.[9]

Critical accounts of Fechner's legacy have painted a nefarious picture of his influence on scientific and cultural history in Europe, particularly in the nineteenth century. Jonathan Crary has argued, for example, that Fechner's quantification of sensation "render[ed] a perceiver manageable, predictable, productive" and made the human body "compatible with arrangements of power."[10] In making these claims, Crary draws on Michel Foucault's theory of biopolitics, which describes the nineteenth century as the historical moment in which life (*bios*) and its processes are incorporated "into the order of knowledge and power, into the sphere of political techniques."[11] Although Crary's valuable study covers extensive ground, it passes over the work of Francis Galton, who did more than any other nineteenth-century scientist to usher Britain into a new biopolitical age using Fechner's theories and experimental techniques.

This chapter begins by establishing Galton's significance in the history of nineteenth-century biopolitics, which depended upon his knowledge of and experiments in psychophysics. As I demonstrate, Galton used Fechner's work to design new instruments that could measure the relative sensitivity of an entire population, making the bodies and sensory processes of people knowable, comparable, and "compatible with arrangements of power." Indeed, Galton's collection of biological data, which included intimate details about the sensory capacities of thousands of people, solidified his eugenic interests and his belief that blindness, deafness, and other disabilities would weaken the national "stock." After discussing the ominous implications of Galton's early work in psychophysics, however, I trace the surprising turns his research took when he began to lose his hearing. As Galton reached the point of near-deafness during the 1890s, I argue, he turned his attention to a subject in Fechner's *Elemente* that he had never explored before: the shadowy world of "below-threshold stimuli."[12] These stimuli acted upon the nerves, as Fechner observed, but failed to reach the threshold required to produce a complete sensation. Inspired by Fechner's theory of below-threshold stimuli, Galton speculated that faint sensations—such as barely audible sounds—could be supplemented by the imagination, and he turned to the poetry of Wordsworth and Tennyson to substantiate his speculations. In so doing, Galton simultaneously opened new lines of scientific inquiry and shed new light on the sensory poetics of Wordsworth and Tennyson.

In this chapter, I consider how Galton's experience of deafness altered his interpretation of literature, science, and the body. He began to reconceive the body's sensory thresholds as something other than hard limits that mark the fixed range of a person's ability; he began to see them as boundaries to be surpassed by the powers of imagination. At the same time, his personal struggle with deafness rerouted his psychophysical research away from eugenics and toward more productive and positive ends. And finally, I show how Galton's experience of hearing loss inspired new interpretations of familiar poems that affirmed the vital powers of imagination. Of course, these revelations neither overshadow nor mitigate Galton's eugenic fantasies or his belief that congenital deafness was dangerously dysgenic. Rather, they prove that deafness could be a source of insight, ingenuity, and creative interpretation rather than a disabling and degenerative condition that endangered national welfare—even for Galton, who spent most of his life trying to prove otherwise.

Fechner's Elemente *in England*

In 1872, twelve years after *Elemente der Psychophysik* first appeared in Germany, the English psychologist James Sully introduced Gustav Fechner's book to the British public in an extensive article in the *Westminster Review* entitled "Recent Experiments with the Senses."[13] Along with Fechner's *Elemente,* Sully discussed the work of Helmholtz, Volkmann, and Wundt, all of whom researched the "phenomena of sensation," or what Sully called the "borderland of Physiology and Psychology."[14] Sully noted that sensation eluded objective inquiry because it was an "individual and subjective" experience.[15] But he observed that Fechner and other German scientists managed to mitigate the "oscillation of individual feeling" by "varying the experiments" and by "taking different states of the same individual, as well as many different individuals" into consideration.[16] This was a crucial point for British scientists who valued objectivity and understood it to be the guiding principle of scientific inquiry, as George Levine, Lorraine Daston, Peter Galison, and others have demonstrated.[17]

From the moment Fechner's work appeared before British readers, then, it was framed as an innovative empirical science that took subjective experience into account. It offered a reliable way to remake specific aspects of interiority available for objective study, a desire that many Victorian scientists shared. As William A. Cohen has observed, "Much of Victorian mental science focuses on differentiating interior from exterior states and on the links between physical and immaterial components of human psychology."[18] Moreover, Fechner's work dovetailed with the physiological psychology that Alexander Bain, William Carpenter, and Thomas Laycock had been writing about since the 1850s. Thus when Francis Galton praised the "new science" of psychophysics in an 1875 letter to a friend, he remarked that although the book was fifteen years old, "the reading world is only now prepared to recognize its merits."[19] Thus he declared that he would be "heartily glad if an English publisher were to bring his work out in translation, believing that it would interest many scientific men and introduce a new and much needed branch of scientific investigation into England."[20] Although Galton's wish would not be fulfilled for almost a hundred years, he gauged the book's value and significance accurately—it did interest a number of scientists, including George Henry Lewes, Henry Maudsley, and others.

By his own account, Galton was most impressed by the central aim of Fechner's book, which was "to show that one fundamental law connects the amount of *sensation* (in the widest sense of the word) with the magnitude of the *exciting cause*," and he concluded that the experiments Fechner used to discern that fundamental law were "most delicate and ingenious."[21] Indeed, his meticulous experiments revealed that physical stimuli must reach a "threshold" (die *Schwelle)* to register on a person's consciousness. Then, building on the work of Ernst Weber, a German physician who described the minimum recognizable difference between two stimuli as the "just-noticeable difference" (JND), Fechner conducted further experiments that involved slowly increasing the intensity of a stimulus (the heaviness of a weight, for example) and asking the subject to continuously evaluate whether or not he or she perceived any differences. After collecting and analyzing the data, Fechner refined Weber's hypothesis by noting that the intensity of a stimulus must increase exponentially for a person to perceive an increase in its intensity. He then developed a logarithm to describe the exponential relationship between stimulus and perception.[22] The Weber-Fechner formula, as it is now known, expresses the "fundamental law" that "connect[s] the amount of *sensation* . . . with the magnitude of the *exciting cause*," and it is this law that so impressed Galton.

While Galton, Sully, and other British scientists marveled at Fechner's work and the formula he derived through rigorous experimentation, other Victorian writers were exasperated by his attempt to describe the relationship between sensation and stimulus in mathematical terms. In *Studies in the History of the Renaissance* (1873), for example, Walter Pater wrote fondly of Renaissance science, which was "all divination, clairvoyance, unsubjected to our exact modern formulas, seeking in an instant of vision to concentrate a thousand experiences."[23] Pater was likely referring to the Weber-Fechner formula, which Sully had introduced to British audiences in the pages of the *Westminster Review* just one year before Pater's book appeared in print. And indeed, Fechner did measure and quantify the "thousand experiences" that constitute a sensation, yielding a totalizing formula that described precisely how sensory experience corresponded to physical stimuli. Given Pater's devotion to the British aesthetic movement, he would have been even more distraught if he had lived to see the rise of experimental aesthetics in the twentieth century, which made ample use of the Weber-Fechner formula to calculate the optimal stimulatory intensities for aesthetic experience.

Fechner's quantification of sensory experience had more pernicious effects on the English populace than Pater intuited, however. As Jonathan Crary argues in *Techniques of the Observer*, the circulation of Fechner's work in the nineteenth century marked the moment when the human subject, "through knowledge of the body and its modes of functioning, was made compatible with arrangements of power." Bringing Foucault's writing on biopolitics to bear on his analysis of nineteenth-century visual technologies, Crary insists that Fechner's "arithmetical homogenization" of the senses "render[ed] a perceiver manageable, predictable, productive."[24] In this model, qualitative differences in perception are obliterated and only quantitative distinctions among perceptual capacities are available for appraisal. Fechner's work formalized the plurality of human difference into rigid hierarchies by translating it into quantified bundles, Crary argues, and enforced a more rigorous distinction between humans and nonhumans (only the former can describe, and therefore quantify, the intensity of their sensations).

Crary analyzes the effects of Fechner's quantification of sensory perception and makes an especially important argument about the new technologies that used psychophysics to make the human body more fully available to market capitalism. Moreover, Crary understands how psychophysics contributed to the production of subjects who have become complicit in the surveillance of their own senses. And yet he does not consider how the concepts in *Elemente* arrived in England or how British scientists like Francis Galton translated Fechner's vision into a biopolitical reality. As Thomas Lemke points out, the "objects of biopolitics" are the biological features of individual people "measured and aggregated on the level of populations," and the collation of data about such features makes it possible to define norms, establish standards, and determine average values."[25] By designing instruments that could measure the sensory capacities of the English populous, Galton used psychophysics to gather data about and produce a statistical map of the sensory capacities of a large population of living bodies. But Galton did far more than make those bodies available to market capitalism or even governmental power, as Crary argues. Rather, Galton offered a statistical representation of their sensory capacities to justify his eugenic theory and make it seem scientifically legitimate.

Galton and Fechner

Psychophysics was especially appealing to Galton, an "apostle of quantification" who became increasingly interested in statistical analysis.[26] After learning of Fechner's ingenious experiments in James Sully's essay in the *Westminster Review* and reading *Elemente* on his own, Galton began to devise ways of making psychophysics "suitable for other applications."[27] According to his biographer, Karl Pearson, Galton's "mind was turning from physical to psychical anthropometry" in 1877 when he gave an address to the Department of Anthropology of the British Association for the Advancement of Science (BAAS).[28] At the very beginning of the speech, Galton made a special point of describing the new lines of inquiry that Fechner's work in psychophysics made possible:

> What . . . I especially wish to point out is, that it has of late years become possible to pursue an inquiry into certain fundamental qualities of the mind by the aid of exact measurements. Most of you are aware of the recent progress of what has been termed Psycho-physics, or the science of subjecting mental processes to physical measurements and to physical laws. I do not now propose to speak of the laws that have been deduced, such as that which is known by the name of Fechner . . . ; but I will briefly allude to a few instances of measurement of mental processes. . . . They will show, what I desire to lay stress upon, that the very foundations of the differences between the mental qualities of man and man admit to being gauged.[29]

Galton tried to impress upon his audience what he believed to be true: that Fechner's pioneering experiments made it possible to take "exact measurements" of a person's mental qualities and to compare those of one man to another to gauge their relative strengths and weaknesses. And although Galton emphasized the measurement of "mental processes" in his first public address on the new science of psychophysics, he integrated Fechner's work into other areas of research, including his work on composite photographs and, as we shall see, his study of sensation.[30]

Shortly after he delivered his 1877 address, Galton began to prepare the ground for his own psychophysical research into the sensory capacities of large populations, which would occupy him over the next decade. Using Fechner's theory of sensory thresholds, he designed a series of instruments and devices that could measure the power of a person's senses, thereby allowing Galton to compare them to the power of another. One of the most important instruments he devised was a small whistle that could ascertain the "upper limits of audible sound in different persons."[31] According to Galton, he conducted "amusing experiment[s]" on "some rather elderly and self-satisfied personages," which revealed "a remarkable falling off in the power of hearing high notes as age advanced."[32] The instrument (now referred to as the Galton whistle) was instrumental in the diagnosis of presbycusis, the loss of hearing in old age, which was precisely the condition that resulted in his profound deafness in the final years of his life.[33]

Throughout the 1880s, Galton used Fechner's theory to design various instruments that would gather sensory data from broad swaths of the English population. As a result of Galton's efforts, psychophysics became an essential part of the broad, scientific effort to collect biological data about the bodies of the English citizenry. In 1880, Galton was appointed chairman of the Anthropometric Committee of the BAAS, which was to orchestrate "The Systematic Examination of Heights, Weights, &c., of Human Beings in the British Empire."[34] Although the committee had collected scores of measurements from different areas in Britain—not quite from the whole empire, as they promised—Galton wanted a more expansive set of measurements, along with fingerprints and photographs of those who were measured. By the time the Anthropometric Committee disbanded in 1885, Galton had opened his own anthropometrics laboratory at the International Health Exhibition in London. Tucked away among the many exhibits at the "Heatheries" was Galton's laboratory, "a compartment only 6 feet wide and 36 feet long, [in which] about ninety persons were measured daily in an elaborate manner."[35] The laboratory was fitted with a new set of instruments intended to measure "keenness of sight; colour sense; judgment of eye; hearing; highest audible note; breathing power; strength of pull and squeeze; swiftness of blow; span of arms; height, standing and sitting; and weight."[36]

When the exhibition closed down a year after it opened, Galton was permitted to move his laboratory to the Science Museum at South Kensington, where it would remain active for another six years. Over the years, he had collected measurements of the sizes, strengths, and perceptual capacities of 9,337 people.[37] In Galton's laboratories, the general public gave their bodies over to measurement, treating the instruments as entertainments in a health exhibition. While they enjoyed pleasurable games or interesting activities the scientist had constructed for them, they unwittingly participated in a large-scale collection of their biological data. Galton would spend years converting the data into a statistical snapshot of a whole population—a graph of intensities, a schematization of the formerly intangible, inaccessible qualities and capacities of British bodies. Galton's anthropometric laboratories carried out one of the first technologically sophisticated and centrally organized attempts to collect vital information about the sensory and physical capacities of the British public and render that information in statistical terms. His laboratories accordingly became a crucial component of Victorian biopolitics, which Foucault describes as the "numerous and diverse techniques for achieving the subjugation of bodies and the control of populations."[38]

Galton's massive collection of data at the Anthropometrics Laboratory allowed him to generate theories about sensation, which he presented in *Inquiries into Human Faculty* (1883). Using Fechner's language, he concluded that "morbidly sensitive persons" were "induced by lower stimuli than . . . the healthy, but the number of just perceptible grades of sensation between them is not necessarily different."[39] Here he reaches the surprising conclusion that the "women of delicate nerves" who suffer from pathological supersensitivity do not have "acute powers of discrimination." He also makes the counterintuitive claim that men, as a rule, "have more delicate powers of discrimination than women."[40] Women lose on both ends of Galton's sensory study: they are too sensitive to be rational, but not sensitive enough to be discriminating. He grouped workers, "idiots," "savages," and the blind together with women, arguing that they were all less sensitive and discriminating than English gentlemen. And to make matter worse, Galton claimed that his experiments confirmed what he expected to find: that the most sensitive people were also "intellectually ablest" and, as he would later argue, the best suited for reproduction.[41]

During this phase of his career, Galton used psychophysics for more ominous purposes than the comparatively benign collection of biological data for governance or management of a large population: he used the science to statistically justify his theory of eugenics. Anticipating his later work in eugenics, Galton began to express in *Inquiries* his concern about the health of "Our human civilised stock," insisting that it is "far more weakly through congenital imperfection than that of any other species of animals."[42] The sources of such imperfection, according to Galton, are the "weakly and misshapen individuals" one encounters on the street. To ensure that his readers grasp the threat that disability poses, he reminds them that "the worst cases are out of sight" and argues that "we should parade before our mind's eye the inmates of the lunatic, idiot, and pauper asylums, the prisoners, the patients in hospitals, the sufferers at home, the crippled, and the congenitally blind."[43] Disabled bodies serve as props in Galton's early rhetoric on eugenics; they are not people to him, but empty signs of the degeneration that would certainly come if marriage and reproduction were not managed with extreme care.

Galton and the Auditory Imagination

In *Inquiries into Human Faculty*, Galton's derision of physical disability emphasizes blindness, but he soon turned his attention to deafness.[44] In an 1885 essay entitled "Hereditary Deafness," which appeared just three years after *Inquiries*, he expressed his serious concerns about the "deaf-mute" communities forming in the United States, which he had read about in an article by Alexander Graham Bell.[45] Galton shared Bell's fears about evolutionary futurity and believed that deaf communities were forming a "marked and degenerate variety of mankind" perpetuated by intermarriage. He insisted that "strong social, and possibly legislative, agencies" would arise to prevent any marital "unions" that were likely to produce "heredity effects harmful to the nation."[46] He also argued that "gesture-languages"—or signing—should be suppressed and that "the philanthropic custom of massing the deaf and dumb together in separate societies, and of making their life as happy as possible in those societies" should be discouraged.[47]

As Galton's hearing began to decline in the 1890s, however, the tone of his writing about the deaf softened considerably. In 1907, at the age of eighty-five, a nearly deaf Galton replied to a letter from Charles Darwin's son, George, soliciting donations for the blind: "I fully sympathise," Galton responded, "and gladly send £2 to help it. But my strongest sympathy is with the deaf. Had I a fairy godmother," he continued, "I would petition that every experimental physicist should be made as deaf as I am, until they had discovered a good ear trumpet, and then that as many fairy-gifts should be heaped on the discoverer as should exceed all he could desire, as well as the thanks and gratitude of all whom he had relieved!"[48] In part, Galton perceived his deafness as a disabling condition, a burden to be relieved by a mechanical device, the "ear trumpet" he desired. But when Galton wished that "every experimental physicist" could be as deaf as he was, he suggested that deafness could be productive and enabling—not because it would cultivate sympathy in the able-bodied, but because it would supply meaningful motives for new scientific inquiry. For Galton, deafness was both a troublesome burden and an affirming potentiality, especially for scientists working on technologies that amplify the senses.

In an 1893 lecture at the Royal Institution, which announced its debt to Fechner in its title, "The Just-Perceptible Difference," Galton began to describe what he called the "auditory imagination"—a faculty that we all develop when we read silently. In fact, he defined reading as "the auditory presentation of the words that are perused by the eye."[49] Without this faculty, he claimed, "It would be . . . impossible to realise the sonorous flow of the passages, whether in prose or poetry, that are read only with the eyes." By virtue of his deafness, Galton had become quite adept at 'realizing the sonorous flow' of prose and poetry, of 'listening' to words printed in text by translating them into imagined sound. In fact, he described his deafness as the enabling experience—one of "two helpful conditions"—that permitted him to "measure the force" of his own auditory imagination; the other is printed copies of the public lectures he attended, which were sometimes distributed in advance.[50] At these lectures, he made a habit of comparing his "capabilities of following the reader when [he is] using [his eyes], and when he is not"—and he found self-reflexive practice to be "a never-flagging source of diversion." Galton took pleasure in exploring the "potency" of his auditory imagination and in hearing through means other than his physiological ear. And his pleasure and his ability to 'follow the lecturer' dissipated when he ceased to experiment in this way: "Should I raise my eyes from the copy," he wrote, "nothing whatever . . . can be understood, the overtones by which words are distinguished being too faint to be heard."[51] In these fascinating experiments, Galton inverted the structuring phonocentrism that, according to Jacques Derrida, privileges the speaking voice over the written word. In deafness, he had become adept at 'hearing' texts and listening with faculties other than those associated with the ear. Moreover, he claimed that we "all . . . cultivate this form of auditory imagination . . . when we are listening to the words of a reader while our eyes are simultaneously perusing a copy of the book from which he is reading."[52]

To develop his theory of the universal auditory imagination, Galton turned to Fechner's psychophysics, but approached it from a more intriguing angle this time: rather than devising instruments that could pinpoint sensory thresholds with accuracy, he began to wonder about the ghostly stimuli that fell below the threshold of perception. George Henry Lewes wrote of these stimuli eloquently in the second volume of *Problems of Life and Mind* (1879), using Fechner's threshold theory to explain how they can affect the nervous system without registering as a complete sensation:

> There can be no sensation without adequate stimulation, and no stimulation without external stimulus. But the contact of a stimulus with a sensitive surface does not suffice for Sensation: it must have a certain energy to disturb the neural equilibrium, and produce an

> excitation; further, that excitation must reach a certain level of relative intensity to produce a change in the state of consciousness.[53]

In this passage, Lewes illuminates Fechner's most profound, but often overlooked, insight: that not all stimuli result in a complete sensation. As Lewes explains, only those that have "a certain energy" may "disturb neural equilibrium" enough to "produce an excitation," and even then, the excitation had to be intense enough to "produce a change in consciousness." What about the stimuli that excite the nervous system, but do not produce any change in consciousness? According to Fechner's theory, these stimuli either dissolve into unconsciousness or, as Nicholas Dames puts it, remain below threshold until they could "accumulate sufficiently" and "burst suddenly upon the consciousness."[54]

Galton took a profound interest in these kinds of stimuli as his hearing faded, so much so that in his 1893 lecture on the auditory imagination, he explained that he wanted nothing more than to move "beyond . . . the frontier of the mysterious region of mental operations which are not vivid enough to rise above the threshold of consciousness."[55] He would not accept that a weak stimulus must remain below the threshold of consciousness, as it would in a person whose hearing capacity had diminished. Rather, he insisted that the imagination could help these below-threshold stimuli rise to consciousness: the imagination, he explained, "originate[s] . . . what may be termed *incomplete* sensations" and "[when] one of these concurs with a real sensation of the same kind, it would swell its volume."[56]

Galton and the Poets

To develop the idea that the imagination could produce "faint sensations" that supplemented the body's physiological circuitry and help bring weak stimuli to threshold, he turned to Wordsworth and Tennyson. Like many Victorian scientists who, according to Gillian Beer, "habitually seamed their sentences with literary allusion and incorporated literature into the argumentative structures of their work," Galton carefully selected the lines that could illustrate, exemplify, or prove his point about sensation.[57] Moreover, he believed that poetry was the appropriate place to search for evidence that the imagination could supplement physical sensation, since "the force of the imagination may endure with extraordinary power and be cherished by persons of poetic temperament." He therefore turned to Wordsworth's "Ode: Intimations of Immortality," seizing on the famous lines in stanza 9 that, according to Galton, "long puzzled his readers":

> Not for these I raise
> The song of thanks and praise,
> But for those obstinate questionings
> Of sense and outward things,
> Fallings from us, vanishings, &c.[58]

Galton overlooks many of the nuances of this passage, disregarding the speaker's dim recollections of childhood and the celestial plenitude he has lost, and reads it instead as an allegory about sensory thresholds. He arrives at this interpretation by referring, in perfect literary-critical fashion, to Wordsworth's marginal notes: "The explanation," Galton tells us,

> is now to be found in a note by Wordsworth himself, prefixed to the ode in Knight's edition. Wordsworth there writes—"I was often unable to think of external things as having external existence, and I communed with all I saw as something not apart from, but inherent in, my own immaterial nature. Many times while going to school have I grasped at a wall or tree to recal [*sic*] myself from this abyss of idealism to the reality. At that time I was afraid of such processes. In later times I have deplored, as we all have reason to do, a subjugation of an opposite character, and have rejoiced over the remembrances, as is expressed in the lines 'Obstinate questionings,' &c."[59]

Galton uses biographical detail and marginal notes to reconstruct the poet's experience and make sense of the lines that Helen Vendler insists are the "heart of the poem."[60] As Vendler argues, the lines about misgivings and questionings figure criticality as compensation for the lost splendor of childhood. And while these are "unpleasant experiences, inexplicable disorientations in a shadowy universe," she argues, they are nonetheless the foundation on which we build "our later trust" in the inward reality of feeling and intellect.[61]

For Galton, however, the poet's "obstinate questionings" of "sense and outward things" refer not to the gift of criticality, but to the specific disbelief that perception is merely the act of sensing the object world. In Wordsworth, he finds proof that sensation is not always the result of a physical encounter, a contact between the sensing body and the external world. He finds an opportunity to read against physical reality and identify an alternate route to sensation that does not involve the physiological circuits ordinarily associated with sensory perception. Wordsworth's inability to "think of external things as having external existence," and his idea that what he communed with was "inherent" in his own "immaterial nature," both support the theory that sensations can originate within the mind. Moreover, they suggest to Galton that the mind can supply a stimulus that augments or supersedes physical stimuli—"external things"—and that perception requires no object, so long as one's powers of imagination are intact.

Galton's literal reading of the Immortality Ode, and his insistence on pairing it with Wordsworth's anecdotal account of an unusual sensory experience, is an example of what Brian Massumi describes as interdisciplinary "poaching." Poaching happens, according to Massumi, when "a concept [is] severed from the system of connections from which it is drawn and plopped into a new and open environment where it suffers an exemplary kind of creative violence."[62] Galton's quotation of Wordsworth's Ode is creative violence, par excellence, because it violates the poem's rhythms, disrupts its flows, and forces unwelcome breaks; it even places a rude ampersand where Wordsworth's lines continue on: "Blank misgivings of a Creature / Moving about in worlds not realised." But, as Massumi points out, this act of creative violence "is only half the story. . . . When you uproot a concept from its network of systemic connections with other concepts you still have its *connectibility* . . . the concept carries a certain residue of activity from its former role."[63] Whenever a concept is poached, Massumi tells us, it carries its affects from the original environment; the poem's affects carry over. In this case, the poem's oscillation between overwrought lament over "The things which I have seen I now can see no more" and sober discovery of a nourishing criticality fuse into Galton's writing about hearing loss and his development of coping mechanisms and new routes to a different kind of fulfillment.

Following his discussion of Wordsworth, Galton turns to Tennyson's "The Holy Grail," a poem about the Knights of the Round Table and their search for a divine and ever-disappearing object. Tennyson famously referred to "The Holy Grail" as "one of the most imaginative of [his] poems" and claimed that it "expressed . . . [his] strong feeling as to the Reality of the Unseen." This Idyll is told in flashback by Percivale in old age to his fellow monk, Ambrosius. Percivale explains that his sister, a Holy Nun, first beheld the Grail and that a vision of the covered Grail appeared before the knights, too, while Arthur was away tending to the needs of the secular world. Percivale, inspired by his sister's fervency, swears that he will quest for it a year and a day, and Galahad, Lancelot, Gawain, and Bors follow suit. When Arthur returns, he is dismayed and makes predictions about the knights who embark on the quest to see the Grail. Ultimately, the knights experience only what they are capable of experiencing: they, according to Arthur, "have seen according to their sight"—which suggests that one's capacity to apprehend an ideal is relative; that is, visions are only as powerful or as real as one's beliefs, dispositions, and faith.

Galton was interested in a passage that appears at the very end of the poem, after King Arthur has heard each of the knights describe his visionary experience of the Grail, or lack thereof. Galton includes these in his 1893 lecture:

> Let visions of the night or of the day
> Come, as they will; and many a time they come
> Until this earth he walks on seems not earth,
> This light that strikes his eyeball is not light,
> The air that smites his forehead is not air,
> But vision, &c.[64]

Arthur's speech is tangled and complicated, and its possible meanings would require a separate essay. But Galton was not interested in the meaning of Arthur's speech within its own textual environment; instead, he was intrigued by the idea, hinted at in the passage, that sensation does not necessarily originate in contact between the external world and the physiological senses. The paradoxical idea that the "light that strikes the eyeball is not light" becomes a meaningful sign for Galton; extracted from its own internal semantic and symbolic network and resituated in his lecture, it becomes evidence that the mind and the material world collude to produce sensations.

Again, Galton violates the aesthetic integrity of the poem, cutting off the final lines of the passage, which Tennyson referred to as "the (spiritually) central lines of the Idylls."[65] But he does so to find a new meaning in it—a meaning that speaks to his experience of deafness and his ongoing search for a theory of supplemental sensations that originate in the mind or cross over from other senses rather than traveling the ordinary neurophysiological routes. And again, Galton reads the poem alongside an anecdote that Tennyson's friend and cofounder of the Metaphysical Society, James Knowles, published in an essay on the late poet in

1892: "Sometimes," Tennyson reportedly said to Knowles, "as I sit alone in this great room I get carried away, out of sense and body, and rapt into mere existence, till the accidental touch or movement of one of my own fingers is like a great shock and blow, and brings the body back with a terrible start."[66] The image of a rapturously disembodied Tennyson was both striking and provocative to Galton, who avoided the common interpretation of these incidents as trance states or as seizures. Rather, he took it as evidence that "the imagination is sufficiently intense to mimic a real sensation."[67]

Conclusion

Although Galton's commitment to eugenics never waned, his latter-day experiments with psychophysics helped to defamiliarize what Lennard Davis describes as "one of the foundational ableist myths of our society"—that speaking and hearing are "the norm."[68] During the 1880s, Galton used Fechner's work to statistically define that norm by collecting biological data from thousands of visitors who passed through his Anthropometrics Laboratory. But he later used different aspects of the science to discover a way around the standard auditory mechanisms of other, less traveled routes of sensory experience. For Galton, those less traveled routes involved supplementation with other senses (reading along while listening) or with the purely creative powers of imagination, at least during the 1890s. But one tantalizing account suggests that he may have purchased an electronic hearing aid at the very end of his life. In January 1911, a close friend of Galton's wrote to the *Times* to share details about the belated scientist's private life and character with the public. He begins with a story about Galton's debilitating deafness and the new device he acquired to alleviate the condition:

> His first sore trial was his deafness, which cut him off from scientific gatherings where at one time he was a familiar figure. This defect he remedied with the help of an electrical instrument very much in the form of a camera. I well remember going to see him a day or two after this new acquisition. Pointing to it as it stood on the table by his side, he said:—"That is my ear. If you will speak to it without raising your voice I shall hear all you say." The experiment was successful and he talked gaily on De Quetelet's letters and digressions from curve of frequency.[69]

Galton seems to have embraced the electronic hearing aid, which restored his hearing by amplifying the physical stimulus so that it reached the requisite threshold. The writer describes the instrument as "remedy" for Galton's deafness, a stubborn "defect" that closed him off from the public life he once enjoyed as a prominent and prolific scientist. And yet, as I have shown, Galton's response to deafness suggests that his loss of hearing was profoundly enabling and productive. Forced to relinquish the soundscapes he had always known, Galton sought to perceive the world in new ways and supplemented what he heard with input from other senses; he moved beyond the biopolitical impulses that guided his earlier work on psychophysics and entertained creative new ideas about below-threshold stimuli; he developed provocative readings of poems by Wordsworth and Tennyson in his quest to prove to himself and to the public that below-threshold stimuli could be supplemented and brought to sensorial fruition; and he even wished that other scientists might be as deaf as he because he saw that deafness was neither a punishment nor a defect, but a powerful source of motivation and an impetus to reconceive literature, science, and the body in ways he never would have bothered to imagine before.

Notes

1. Gustav Theodor Fechner, *Elements of Psychophysics*, trans. Helmut E. Adler, vol. 1 (New York: Holt, Rinehart, and Winston, 1966), 1. Unless otherwise noted, all references to Fechner's *Elements* will refer to this edition. The original book, *Elemente der Psychophysik* (Leipzig: Breitkopf and Hartel, 1860), appeared in two volumes in German. Only the first volume has been translated into English.

2. Ibid., 5. According to Michael Heidelberger, Fechner does not fall into the trap of naïve materialism; rather, he argues, Fechner's materialism is nonreductive "because it describes life and consciousness as having an independent, original nature that cannot be further reduced to physical phenomena." See Michael Heidelberger, *Nature from Within: Gustav Theodor Fechner and His Psychophysical Worldview* (Pittsburgh: University of Pittsburgh Press, 2004), 73.

3. Fechner, *Elements of Psychophysics*, 1. The *OED* dates the first usage of "psychophysics" in English to a July 1875 issue of the *North American Review*. This attribution is inaccurate, however. The first usage of the term appeared in Emil Du Bois-Reymond's essay "The Limits of Our Knowledge of Nature," *Popular Science Monthly*, May 1874, 27. The first usage in British sources appears to be a selection of "Critical Notices" by R. Flint in *Mind* 1 (1876): 117; the same volume featured an article on Fechner and other psychologists by James Sully, who first introduced *Elemente der Psychophysik* to the English-speaking world in a review essay entitled "Recent Experiments with the Senses," *Westminster Review* 98 (July 1872): 165–98. Sully never used the word "psychophysics" in his 1872 review; he referred instead to the "psycho-physical law" that Fechner proposed.

4. Sully, "Recent Experiments," 178.

5. Galton to Mrs. Hertz, June 4, 1875, in Karl Pearson, *The Life, Letters and Labours of Francis Galton*, vol. 3B: *Characterization, Especially by Letters, & Index* (Cambridge: Cambridge University Press, 1914–30), 464.

6. William James, *The Principles of Psychology* (New York: Henry Holt, 1890), 1:534.

7. Ibid., 546. James insisted that Fechner's psychophysical experiments were "peculiarly fragile" because they depended on the idiosyncratic judgments of test subjects and on psychologists' accounts of those judgments.

8. George A. Gescheider, *Psychophysics: The Fundamentals*, 3rd ed. (Mahwah, NJ: Lawrence Erlbaum Associates, 1997), ix.

9. Jay Hetrick, "Aisthesis in Radical Empiricism: Gustav Fechner's Psychophysics and Experimental Aesthetics," *Proceedings of the European Society for Aesthetics* 3 (2011): 140.

10. Jonathan Crary, *Techniques of the Observer: On Vision and Modernity in the Nineteenth Century* (Cambridge: MIT Press, 1992), 147.

11. Michel Foucault, *The History of Sexuality*, vol. 1: *An Introduction*, trans. Robert Hurley (New York: Vintage Books, 1978), 141–42.

12. Nicholas Dames discusses Fechner's theory of below-threshold stimuli and its relevance to nineteenth-century reading practices in *The Physiology of the Novel*, but Galton's interest in these stimuli pertained to auditory experience rather than reading. It is no surprise, then, that Galton does not appear in Dames's study. For his discussion of below-threshold stimuli, see Nicholas Dames, *The Physiology of the Novel: Reading, Neural Science, and the Form of Victorian Fiction* (Oxford: Oxford University Press, 2007), 201–2.

13. David Burbridge, "Galton's 100: An Exploration of Francis Galton's Imagery Studies," *British Journal for the History of Science* 27, no. 4 (1994): 445 n. 18. Burbridge notes that Sully's book *Sensation and Intuition* (1874) also helped distribute Fechner's work to British audiences.

14. Sully, "Recent Experiments," 165.

15. Ibid., 167.

16. Ibid.

17. George Levine, *Dying to Know: Scientific Epistemology and Narrative in Victorian England* (Chicago: University of Chicago Press, 2002); Lorraine J. Daston and Peter Galison, *Objectivity* (New York: Zone Books, 2007).

18. William A. Cohen, *Embodied: Victorian Literature and the Senses* (Minneapolis: University of Minnesota Press, 2008), 5.

19. Galton to Mrs. Hertz, in *Characterization*, 464.

20. Ibid.

21. Ibid.

22. Nicholas Dames offers a helpful definition and discussion of the Weber-Fechner formula; see *Physiology of the Novel*, 179–82.

23. Walter Pater, *The Renaissance: Studies in Art and Poetry. The 1893 Text*, ed. Donald L. Hill (Berkeley: University of California Press, 1980), 83.

24. Crary, *Techniques of the Observer*, 147.

25. Thomas Lemke, *Biopolitics: An Advanced Introduction*, trans. Eric Frederick Trump (New York: New York University Press, 2011), 5.

26. Stephen Jay Gould, *The Mismeasure of Man: The Definitive Refutation to the Argument of "The Bell Curve"* (New York: Norton, 1981), 107.

27. Galton to Mrs. Hertz, in *Characterization*, 464.

28. Karl Pearson, *The Life, Letters and Labours of Francis Galton*, vol. 2: *Researches of Middle Life* (Cambridge: Cambridge University Press, 1924), 228.

29. Francis Galton, "Address to the Department of Anthropology," in Section II, "Notices and Abstracts," *Report of the Forty-Seventh Meeting of the British Association for the Advancement of Science; Held at Plymouth in August 1877* (London: John Murray, 1878), 95.

30. For excellent work on Galton's use of Fechner's theory of JNDs in his writing on composite photography, see Allan Sekula, "The Body and the Archive," *October* 39 (Winter 1986): 51; and Frans Lundgren, "The Politics of Participation: Francis Galton's Anthropometric Laboratory and the Making of Civic Selves," *British Journal for the History of Science* 46, no. 3 (2013): 461.

31. Francis Galton, *Inquiries into Human Faculty and Its Development* (London: Macmillan, 1883), 26.

32. Ibid.

33. Ironically, Galton made these comments on deafness in 1883, just a few years before his own hearing began to rapidly deteriorate.

34. Francis Galton, "Report of the Anthropometric Committee," in Section I, "Reports on the State of Science," *Report of the Fiftieth Meeting of the British Association for the Advancement of Science; Held at Swansea in August and September 1880* (London: John Murray, 1880), 120.

35. Francis Galton, "On the Anthropometric Laboratory at the Late International Health Exhibition," *Journal of the Anthropological Institute of Great Britain and Ireland* 14 (1885): 206.

36. Ibid., 205.

37. Francis Galton, "Some Results of the Anthropometric Laboratory," *Journal of the Anthropological Institute* 14 (1885): 275.

38. Foucault, *The History of Sexuality*, 140.

39. Galton, *Inquiries into Human Faculty*, 20.

40. Ibid.

41. Ibid.

42. Ibid., 16.

43. Ibid.

44. For a comprehensive discussion of Galton's views on deaf communities, see Jennifer Esmail, "A Deaf Variety of the Human Race?" in *Reading Victorian Deafness: Signs and Sounds in Victorian Literature and Culture* (Athens: Ohio University Press, 2013), 133–62.

45. In his contribution to this collection, James Emmott discusses Alexander Melville Bell's work on Visible Speech, which he used to help deaf children vocalize. Emmott observes that although Bell's approach was utopian in many ways, his son, Alexander Graham Bell, took a dark turn toward eugenics in his aggressive support of the oralist movement.

46. Sir Francis Galton, "Hereditary Deafness," *Nature* 31 (January 1885): 270.

47. Ibid.

48. Galton letter to George Darwin, November 2, 1907, in *Characterization*, 584.

49. Francis Galton, "The Just-Perceptible Difference," *Notices of the Proceedings at the Meetings of the Members of the Royal Institution of Great Britain, with Abstracts of the Discourses Delivered at The Evening Meetings*, XIV, 1893–1895 (London: William Clowes and Sons, 1896), 18.

50. Ibid.
51. Ibid., 19.
52. Ibid., 18.
53. George Henry Lewes, *Problems of Life and Mind*, vol. 2: *Mind as a Function of the Organism*, 3rd series (London: Trübner, 1879), 364.
54. Dames, *Physiology of the Novel*, 181.
55. Galton, "The Just-Perceptible Difference," 13.
56. Ibid., 17.
57. Gillian Beer, *Open Fields: Science in Cultural Encounter* (New York: Oxford University Press, 1996), 174.
58. Galton, "The Just-Perceptible Difference," 16.
59. Ibid.
60. Helen Vendler, "Lionel Trilling and the *Immortality Ode*," *Salmagundi* 41 (1978): 81.
61. Ibid., 83.
62. Brian Massumi, *Parables for the Virtual: Movement, Affect, Sensation* (Durham, NC: Duke University Press, 2002), 20.
63. Ibid.
64. Galton, "The Just-Perceptible Difference," 16.
65. Alfred, Lord Tennyson, *Idylls of the King* (New York: Penguin, 1983), 90.
66. Galton, "The Just-Perceptible Difference," 17.
67. Ibid.
68. Lennard Davis, *Enforcing Normalcy: Disability, Deafness and the Body* (London: Verso, 1995), 15, quoted in Esmail, *Reading Victorian Deafness*, 7.
69. Montague Crackanthorpe, "Sir Francis Galton," *The Times*, January 25, 1911. The device Crackanthorpe is referring to may be the Acousticon, an early electrical hearing aid. This device was developed at the turn of the century by Miller Reese Hutchison, an American inventor and electrical engineer whose earlier invention—the Akouphone (1895)—is widely regarded as the world's first electrical hearing aid. In 1902, Hutchison traveled to London to demonstrate how his new Acousticon worked. According to an account of his visit, he showed the device at Buckingham Palace and at various institutions; Queen Alexandra was so impressed that she awarded him the gold medal for his invention. For this account, see A. L. Griffith, "The Acousticon," *The World To-Day* 5, no. 1 (July 1903): 855. As a Fellow of the Royal Society, Galton certainly would have been aware of Hutchison's demonstrations, especially since they piqued the interest of the queen and won her enthusiastic approval.

CHAPTER 6

Performing Phonographic Physiology

James Emmott

In January and February 1854, the German scholar Chevalier Bunsen convened a group of leading philologists and phoneticians for a series of "Alphabetic Conferences" at his ambassadorial residence in London. The group focused on a difficult linguistic problem that had exercised individuals for centuries: how to identify and arrange human speech sounds in a new symbolic system that would be universally usable across all the languages of the world. The potential of such a system had become increasingly clear for a rapidly globalizing nineteenth century. It would meet a scientific need by offering insight into the sound relations of diverse languages and the mechanism of speech; it would meet a practical need with its promise that a new dawn of intercultural conversation was at hand, galvanized by a parallel explosion in communications technologies that seemed tantalizingly ready to offer intermediary assistance. To achieve this goal, it would first be necessary to move beyond conventional alphabets. It had long been recognized that the relations of spelling to speech were both arbitrary and maddeningly inconsistent—the same sound could be expressed by multiple combinations of letters, and multiple sounds could share identical orthographic formulations. The learned experts gathered for the conferences in London directed their efforts toward the exhaustive collation of the ways in which speech sounds were rendered in existing alphabets, in the hope that these might be refined and reconfigured in novel, complex combinations. Yet for those who were most sensitive to the range of vocal variations within and between languages, it was evident that conventional systems of spelling would not be capable of adequately representing them.

At this historical moment, decades before the appearance of the invention that decisively took the name in 1877, the word "phonography" (literally "sound-writing") designated a variety of practices and techniques that sought to fix and transcribe intangible vocal sounds into recorded forms. The previous quarter-century had seen the gradual development of elocutionism, or vocal training, into what was now becoming called vocal science. The two major influences on the field in this period had issued from the late eighteenth-century writings of Thomas Sheridan and John Walker. Each had promoted different techniques: Sheridan thought that the ideally trained voice was one that advanced toward the purity of natural laws, whereas Walker was committed to identifying the mechanical processes involved in the physiology of vocal production and devising annotations to describe them. The latter method was to inform the so-called mechanical school, which went on to become largely ascendant at midcentury. Among the most prominent of its adherents was the Scottish elocutionist Alexander Melville Bell, the son of Alexander Bell (a well-known vocal teacher in London and the author of a series of popular guides earlier in the century). Melville Bell's own phonetic system, Visible Speech, published in book form in 1867, was one of the major debts of the International Phonetic Alphabet (IPA) in the late 1880s, which, based on physiological principles, marks the consolidation of the mechanical approach that defined modern vocal science.[1]

In this chapter, I argue that Bell's idiosyncratic but highly influential version of physiological phonetics can help us see more clearly how—in the years just before the appearance of Edison's invention—the human body was already being figured in phonographic terms: as an apparatus, increasingly understood mechanically, that records received stimuli and replays them as performed behaviors. In doing so, I show how the fields of physiology, phonetics, and phonography are mutually determined in the 1860s and 1870s, and how each draws on a shared understanding of articulatory performance in accounting for the ways that multiple, sustained impressions and movements record their accumulated traces in the fabric of bodies also furnished with the capacity to replay them. In making the case for connecting vocal physiology with the new meanings of phonography, I examine the development and implementation of Bell's Visible Speech, and then move on to recover two aspects of its broader contextual constellation: first, a sequence of discourses on physiological memory that attended the emergence of the phonograph, and second, the retrospective foil offered by Bernard Shaw's 1912 play *Pygmalion*, whose wry satire on the phonetic culture of the late nineteenth century not only contains a profusion of phonographic metaphors, but also registers some discomfort with the disquietingly implicit agenda of such vocal pedagogy to "humanize" its subjects.

Visible Speech

In *The Principles of Elocution* (1878), appearing in its fourth revised and expanded edition as the phonograph emerged into the consciousness of the world, Alexander Melville Bell maintained that "speech is wholly conventional in its expressiveness, and mechanical in its processes."[2] Elocution, he held, "must embrace the Physiology of Speech—the mechanics of vocalization and articulation," its mastery relying upon learning what Bell called the "principles of Instrumentation." A student "should be made acquainted with the instrument of Speech *as an instrument*, that all its parts may be under his control, as the stops, the keys, the pedals, and the bellows, are subject to the organist."[3] It had taken him some time to reach this position. Following the 1854 conferences, Bell had looked on as a series of attempts toward a "universal alphabetics" had fallen short of ideal completeness. He was an articulate enthusiast of the cause, explaining in a lecture to the Society of Arts in March 1866 that "a system of letters which, when learned in connection with any one language, would be vocalised with uniformity in every other language, has long been felt to be one of the great wants of the world."[4] His approach to the problem was distinctive. He reported that he had alighted on an alternative method, one that avoided the "insuperable obstacles" that had hitherto stalled others' progress. Unlike them, he explained,

> I worked from different data, and by a totally different process. . . . Instead of going to languages to discover the elements of utterance, I went to the apparatus of speech, and, after many partial failures, but with gradual approximations to success, during a long series of years, I had the satisfaction ultimately of discovering, with demonstrable certainty, the complete physiological basis of speech, and of establishing an organic scale of sounds which could not but include all varieties, known and unknown.[5]

In contrast to the effort to collate existing orthographic symbols into a new universal system, Bell's goal was "to discover, *from the organs of speech*, all the modifications of which they were susceptible." From this systematization of the elements of vocal physiology, Bell suggested that "all possible shades of sound might be gathered, and every alphabetic variety in languages might certainly be found and recognized."[6]

Since its symbols were devised as visual analogues of the vocal organs that formed this variety of sounds, Bell named his new alphabet Visible Speech. "All writing may be said to be, in a sense, visible speech; that is, it is a visible record of conventional language," he explained in a lecture on the subject, "but the system of Visible Speech is physiological, and records the actions of the mouth, irrespective of any particular employment of them."[7] Bell's system would record not the arbitrary shapes of traditional letters, whose transformation into spoken utterance depended on often illogical conventions, but rather the "actions of the mouth," in symbols that reflected the physiological arrangement required for a given speech sound—the lips open or closed, the soft palate depressed just so, the passage of air through the nose, and so on. The idea was that any suitably trained person who followed the directions to reproduce those actions would find that the vocal sounds associated with them would follow automatically. "Whatever the mouth can do, you can write," Bell declared, "and whatever you write, any student of the system can read—to whatever language the written matter may belong."[8]

Visible Speech could be described as a universal alphabet because it claimed to enumerate and symbolize every sound, linguistic and nonlinguistic, that the vocal organs themselves were theoretically capable of producing. Where a sound required the simultaneous operation of more than one elementary action, the symbols were gathered into compounds, allowing even the most complicated sounds to be represented. The special distinction of Visible Speech, retained in modern-day phonetics, was therefore that every part of a symbol expresses a physiologically and phonetically meaningful feature, in contrast (for instance) to Roman script, which is littered with redundant, meaningless details (such as the extension of stalks above and below the general line, in *b* and *g*). Visible Speech claimed to hold the universal key to the representation of all vocal sounds. This remarkable aspect of its typology permitted its further description as a "self-interpreting" system, which connected the whole taxonomy of phonetic representations directly to the physiology of the human voice—a symbolically complete metalanguage. It was a system of spelling sound that even in the middle decades of the nineteenth century would have been readily understood as phonographic. The extensive promise of Visible Speech as a technique of sound-writing is reflected in the testimony of a contemporary:

> A full sneeze, for example, is a complex operation: it comes among what are called inarticulate sounds; but Mr Bell writes it down, and, for aught we know, could undertake to furnish every member of the House of Commons with a symbol representative of his own particular sneeze, as distinguished from those of all his colleagues.[9]

In Bell's system, these transcriptions had a dual purpose or possibility, for they enabled not only the capture of vocal phenomena, but also the reproduction (through phonetic rearticulation) of what had been transcribed. Indeed, it was in this sense that Visible Speech moved beyond the symbolic to the properly indexical: its characters did not just represent, they also self-interpreted; they did not merely record the general, but replayed the particular.

The identification of the physiological components of vocal sounds with the mode of their mechanical reproduction had previously only been approximated, with the use of speaking machines—a special variety of a wider tradition of automata dating back several centuries that were contrived to replicate a range of physiological processes, from eating and drinking to the workings of the circulatory system. As Thomas L. Hankins and Robert J. Silverman have shown, speaking machines emerged from the work of phoneticians, elocutionists, and stenographers who were concerned with rendering their investigations more objective. For them, such a machine might "serve as a standard for pronunciation; it could help to analyze speech sounds into their phonetic components; and it could aid in forming a truly phonetic shorthand, because a phonetic sound could correspond to a unique arrangement of [its] elements."[10] The English phonetician Alexander John Ellis was an influential advocate of this approach, observing in his 1845 book *Alphabet of Nature* that "it is impossible that any person in analysing sounds can do more than analyse his own sensations"; since these "may never occur in any other individual," it cannot be done "without the aid of a machine."[11] By contrast, Bell believed that he had indeed distinguished objectively a complete and universal set of dispositions. In describing the human vocal apparatus figuratively as a "speaking machine" in his first published version of *Visible Speech* in 1867, Bell was saying more than he may have realized—for in rendering an actual machine unnecessary for the task, he had figured the body itself as one.[12] The ideal student of Visible Speech commanded the physiological knowledge to replay in bodily actions the sounds that the system had recorded in its phonetic transcript.

Bell used a series of "public experiments in the Writing of Languages" to stage the performance and display of the phonographic capacities he had identified. The demonstrations involved Bell's training of his sons to acquire mastery of the system—accomplished, he claimed, in just a few days—such that they were "enabled to pronounce, at sight, the most difficult and peculiar words that could be selected from the Eastern and other Languages; often involving combinations of sound which the readers had never heard before their own organs gave them utterance."[13] In a published letter to the journal *Reader*, Ellis himself gave an account of one such occasion conducted in London in the spring of 1864. Bell's sons, with no advance knowledge of the details of the challenge about to be posed to them, were sent out of the room, while Ellis "dictated slowly and distinctly the sounds which I wished to be written." As an expert in phonetics, Ellis clearly reveled in devising ingenious ways to test the system's limits. The sounds he dictated to Bell

> consisted of a few words in Latin, pronounced first as at Eton, then as in Italy, and then according to some theoretical notions of how the Latins might have uttered them. Then came some English provincialisms and affected pronunciations, the words "how odd" being given in several distinct ways. Suddenly German provincialisms were introduced. Then discriminations of sounds often confused, as *ees, is'* (Polish), *eesh, ich* (German), *ich* (Dutch), *ich* (Swiss), *ouï, oui* (French), *we* (English), *wie* (German), *vie* (French). Some Arabic, some Cockney-English with an introduced Arabic guttural, some mispronounced Spanish, and a variety of shades of vowels and diphthongs.[14]

Ellis explained his choices by noting that "the sudden changes and the confusion would utterly prevent anyone from guessing by the context," and that the particular distinctions between vowel sounds that he had deployed "would be very difficult either to seize or to imitate except by persons thoroughly used to appreciate [*sic*] such sounds, or led by a strictly physiological system of symbolization to conceive and utter them."[15] Visible Speech was claimed to be just such a system, and according to Ellis's account, it worked. Having transcribed the sounds into the symbols of Visible Speech, Bell recalled his sons to the room, and they articulated them according to the instructions. "The result was perfectly satisfactory," Ellis enthused, noting how "Mr. Bell wrote down my queer and purposely-exaggerated pronunciations and mispronunciations and delicate distinctions in such a manner that his sons, not having heard them, so uttered them as to surprise me by the extremely correct echo of my own voice." He went on: "I was not satisfied with approximations, and I obtained correct imitations. Accent, tone, drawl, brevity, indistinctiveness, were all reproduced with surprising accuracy." This was a range of features that had hitherto eluded conventional methods of sound-writing, and no such methods had been adequately sophisticated as to be able to register and reproduce a vocal phenomenon so phonographically indexical that it could be deemed an "echo." Having studied alphabetic systems for over twenty years, Ellis did not "know of one which could have produced the same results," concluding that "so far, then, as I am able to judge, Mr. Bell has solved the problem." He held that the success of Bell's technique, appropriate to its name, was secured by a physiological fidelity that was practically visible: "I could, as it were, trace the alphabet in the lips of the readers."[16]

There were many other staged demonstrations in which the Bells virtuosically performed the physiological transformation of transcribed input into perfectly articulated output. In a memoir, Alexander Graham Bell (the middle in age of the sons) recalled a public lecture at which members of the audience were "invited to make any sorts of sounds they desired." The volunteers called to his father's platform duly "uttered the most weird and uncanny noises," and the young man rendered them all with exactitude, including an obscure and difficult Sanskrit vowel that he gave correctly without having heard the sound before, and a "curious rasping noise that was utterly unintelligible" to him, but recognized at once by the audience as the sound of sawing wood, "which had been given by an amateur ventriloquist as a test."[17] The method proved so successful that Melville Bell identified it almost immediately as being applicable to the teaching of deaf mutes, which was after all the condition into which his sons had effectively placed themselves in the Visible Speech performances. Since the deaf could not memorize sounds as heard, the physiological basis of Visible Speech was its peculiar advantage: by learning the ways in which the vocal apparatus was configured and utilized in producing articulations, the practical means for rendering the full range of vocal sound could be internalized without necessary auditory reference to the phenomenal sounds themselves. Ellis had actually anticipated this pedagogical development: in a follow-up letter to the

Reader in 1865, he had speculated that the successful promulgation of Bell's scheme would be secured "by transfusing it into living organisms which will give his written symbols motion and meaning."[18]

Graham Bell began to enact this process of transfusion in the teaching that he undertook at Susanna Hull's school for deaf children in London in 1868. In a class with two of these young pupils, his biographer Robert V. Bruce tells us, Bell

> sketched the profile of a face, including the "insides of the mouth" (as he explained to the girls by finger spelling). Then he rubbed out all but the lower lip, the point, front, and back of the tongue, and the glottis. Those curved lines in their respective facings constituted the Visible Speech symbols for "back," "front," "point," "lip," and "voice."[19]

By the end of the first lesson, the girls had learned a dozen sounds. The culmination of the practice, forged in London, followed shortly after the family's move overseas. In 1871 Graham Bell undertook a demonstration of the method before an audience of "influential Educationalists" in Boston, Massachusetts, which he relayed in a letter to his parents. One of his congenitally deaf pupils, Theresa Dudley, who had also been mute for most of her life, "read from the symbols words in German, French and Zulu—introducing clicks." Bell then "invited the audience to dictate words in *any language.* Theresa Dudley did not fail in a solitary instance," he reported. "The best of it," he went on, "is that she does not know yet that she uttered words at all." In a further twist, Bell illustrated how Theresa could "vary the 'timbre' of her voice at will"—as he put it, how she "could inflect it mechanically" under his direction. Following the motions of Bell's hand, another pupil, apparently without quite knowing it, sang first a scale, then a rendition of "God Save the Queen," and then a short extract by the eighteenth-century poet Robert Lloyd.[20]

What is spectacularly revealed in these performances is the Bells' extraordinary conception of the body as a machine for reproducing sound. The theory and practice of their phonetic system augmented transcription with translation, and transformed the static recording of symbols into active rearticulation by "transfusing" into their subjects the capacity for the performance of what I have here called "phonographic physiology." The twin processes of translation and articulation, based on the understanding of the human voice and ear, and made automatic by a self-interpreting symbolic vocabulary, figures this system as phonographic (*avant la lettre*) in the Edisonian—rather than the merely stenographic—sense. In Visible Speech, sounds were not transcribed in order to preserve them for their own sake, or simply to communicate textually. Instead, the system aimed at what amounted to physiological recording and replay. Moreover, as Graham Bell himself keenly observes, his pupils were practically unaware of their vocal achievements. Like his own earlier automatic articulations of previously unheard sounds via the transcribed symbols of Visible Speech, Bell's pupils performed mechanically. Theresa "does not know . . . that she uttered words at all." The device that emerged from Edison's laboratory just a few years later and which appropriated the term "phonography" was to be a machine whose operation was similarly mechanical and unconscious.

Mechanical Memory

The phonograph cylinder worked by revolving while a stylus inscribed whatever acoustic vibrations put it in motion. When the instrument is arranged to replay what it has recorded, its stylus indifferently retraces its path along the grooves it made. Shortly after Edison's invention had emerged from his Menlo Park laboratory, the philosopher Jean-Marie Guyau struck upon the device (as many have similarly struck upon the latest technology) as an apt metaphor for the working of human memory. In citing approvingly the Belgian physiologist Joseph Delboeuf's claim that "the mind is an album of phonographic recordings," Guyau suggested in 1880 that just as the "resonances of the voice are transferred to a needle" in the case of the phonograph,

> it may well be that in a similar way invisible lines are incessantly engraved into the cells of the brain, lines that constitute the beddings for the nervous currents. When, after some time, the current happens to encounter one of these previously formed beds, through which it has already passed before, it engages itself in them once again. Consequently the nervous cells resonate as they did the first time, and this comparable resonance corresponds psychologically to a sensation or thought that is similar to the forgotten sensation or thought.[21]

Guyau's invocation of vibration in the term "resonance" is not accidental. Delboeuf's claim emerged from a broader argument proposing, as Laura Otis puts it, that "memories were changes in patterns of molecular vibration, just as a stone thrown into a pond affects the wave pattern created on the surface by stones that have been thrown in shortly before it."[22] Memories, then, are composite phenomena—"accumulated capital" in Delboeuf's phrase—originating in the vibratory properties widely thought at that time to be fundamental to the nature of physiological matter. This notion of accumulation is foreshadowed in Herbert Spencer's 1863 *Principles of Biology*, in which Spencer claims that all physiological development is a process of combination and compounding, in his molecular focus on how the "mutual play of forces . . . produces a difference in the form which the aggregate of them assumes."[23] Spencer's notion of aggregation is itself drawn implicitly from the composite form of multiple vibrations suggested by the metaphor of competing wave patterns in water, and the same basic principle is found in the phonograph groove, which materially sums the vibratory multiplicity of acoustic phenomena in a single resultant form.

In Delboeuf's and Guyau's post-Edisonian reflections, the phonograph provides for them a new physical analogue for a way of thinking about the connections between physiology, articulation, and memory that already had been in development for some years before Edison. In 1869, for instance, the British physician H. Charlton Bastian had published a provocative series of articles addressing the question of what he called the "physiology of thinking." Bastian's premise is that thoughts are formed and exchanged in language, and that language is fundamentally an act of articulation. "We may ask then," he writes, "whether, in using language as a vehicle for thoughts, words recur or are revived primarily as ideas of sound, or as revived remembrances of articulatory efforts. Are they ever, in fact, primarily revived as 'suppressed articulations'?"[24] For Bastian, an instance of language-as-thought is both an acoustic event and a physiological configuration of the human vocal apparatus. He suggests that what we ultimately bring to mind in this process of mental recurrence or revival is the latter: a memory resides not merely as an abstract datum, but as written into the very fabric of the body. In terms later echoed by Guyau, Bastian describes his account of how sensory impressions "pass along definite routes to certain parts of the cerebral hemispheres."[25] The recollection of these impressions, he held, revives "precisely the same parts of the hemispheres" that had been activated by the original event, and the "same nerve-fibres, and same nerve-cells" are "called into activity as were previously concerned in the perception of the original impression."[26]

This automatic theory of muscular memory was explicitly revived later in the century by one Theodate L. Smith, whose article-length treatise on the subject returns us to the work of Alexander Graham Bell. Writing for the *American Journal of Psychology* in 1896, Smith reports the case of Edith Thomas, a young pupil of Bell's (and sometime classmate of Helen Keller and Laura Bridgman) who had been deaf and blind from the age of four, and who had become increasingly mute since then. When Edith was nine years old, Bell tested her abilities to reproduce movements by means of motor imitation, just as he had done with many other pupils. Smith reports:

> She succeeded fairly well, pronouncing the letter K, which offers peculiar difficulty to deaf mutes, with unusual distinctness. When asked to repeat the letter some hours later, she called with an almost perfect enunciation, "Kitty, Kitty, Kitty." Investigation revealed the fact that when at the age of four years the gradual loss of speech had followed that of sight and hearing, the last intelligible word spoken by the child was "Kitty." The reproduction was unconscious, the child having absolutely no idea of what she had done. It was not, then, a reproduction of the word *as heard or associated with something*, but of a *muscular movement*, which, latent for five years, was recalled by the suggestion of a similar movement. This incident suggested that possibly, under normal conditions, the muscles play a greater part in our memories than we are accustomed to assign to them.[27]

More than thirty years after the original Visible Speech performances and the associated teaching methodology had been devised, Smith here endorses the connections that the Bells had intuited between voice, physiological movement, habit, and memory formation. As he puts it, "Every teacher has observed children busily moving their tongue and lips during the memorizing of a lesson," but they do so, he claims, "without thinking that the movement was not a mere habit, but a real aid in the process of memorizing."[28] In Smith's theoretical model, the physiologically latent store of articulated memories is reactivated by the stimulus of a vocal action that prompts the muscles and neural fibers, stylus-like, to retrace their grooves.

"You Can Turn Her On as Often as You Like"

In her recent book *Heart Beats*, Catherine Robson has identified the seventy or so years from 1875 as the "heyday" of memorization in the pedagogy of poetry in Britain and the United States.[29] That this period closely corresponds to the moment of greatest prominence of the mechanical school of elocutionary teaching, practiced by Bell and others, is surely no coincidence. Neither, as Ashley M. Miller and others have shown, was nineteenth-century prosody in general ever far from understandings of the physiology of memory.[30] These contextual affiliations suggest the importance of a further perspective, unaddressed since the start of this chapter, on what I have so far discussed. For one of the most significant mirrorings between memorized recitation and the quest for a universal alphabetics concerns their shared objective of carefully inculcating (trans)national cultural knowledge by means of physiological transfusion, and the implicitly operative trope of "civilization" in the practice of vocal education. In his 1864 *Reader* article, Alexander John Ellis had expressed his optimism that the emergence of a truly universal alphabet would collapse cultural borders and divisions and "rapidly become a great social and political engine."[31] This was a radical goal enthusiastically shared by the dramatist Bernard Shaw, himself a keen phonetic writing aficionado. His 1912 play *Pygmalion*, to which I will now turn, points toward some of the conflicted consequences of this proposed engine of transformation.

Pygmalion, which Shaw had devised as early as 1897, returns us to the culture of late nineteenth-century phonetic science.[32] The background to the play is well known, but certain aspects of its provenance are worthy of remark in establishing a relation with Melville Bell's Visible Speech. To begin with, Shaw had connections to Bell through the latter's nephew Chichester, who seems to have been responsible for passing to Shaw a copy of his grandfather's 1847 play *The Bride*, in which the valet Allplace is introduced as having been taken into the family of the unremarkable aristocrat Sir Cicero Pandect for the purpose of instructing him in proper manners.[33] The central theme of transformation bears a close resemblance to *Pygmalion*, and is just one signal of the considerable affinities, formal and informal, between Shaw and the Bell family. Bell is named as a "hero" in the preface, and the playwright was surely aware that Melville Bell's wife (who was deaf) was named Eliza. Shaw himself learned Visible Speech in the 1870s, and *Pygmalion*, titled "The Phonetic Play" in the original manuscript, was drafted in Pitman shorthand—the curious concern with phonetics and phonography manifesting itself in both form and content. The male protagonist Higgins (capable, he absurdly claims, of pronouncing 130 distinct vowel sounds) reflects some composite of phoneticians stretching from Bell to Henry Sweet to Daniel

Jones, among others (Sweet's own system, Broad Romic, was another of the crucial developmental steps on the path toward the codification of the IPA).[34] Most significant of all, in *Pygmalion*—a text suffused with the imagery of recording and refashioning—Shaw realizes the process of voice training as a series of performances of writing and reading sound.

In the play, even before the introduction of Eliza, Shaw presents Higgins as someone deeply invested in the performative power of phonetics, but whose interest is marked by extreme emotional detachment. As the play opens, Higgins sits under the shelter of a church portico in Covent Garden, surreptitiously recording the varied diction of the people around him, arousing the suspicions of a crowd. To defend himself, Higgins enthusiastically casts himself in the role of scientific performer. Turning his observations into a form of variety entertainment, he responds to each bystander's spoken objections in turn, gleefully identifying their place of birth: Selsey, Lisson Grove, Hoxton. He is quizzed on his method by an onlooker:

> THE GENTLEMAN [PICKERING]. How do you do it, if I may ask?
>
> THE NOTE TAKER [HIGGINS]. Simply phonetics. The science of speech. Thats my profession: also my hobby. Happy is the man who can make a living by his hobby! You can spot an Irishman or a Yorkshireman by his brogue. I can place any man within six miles. I can place him within two miles in London. Sometimes within two streets.[35]

Shaw figures Higgins as an exemplar of a certain variety of scientist whose obsession with intellectual work precludes authentic engagement with his fellow humans. Shaw describes him as "heartily, even violently interested in everything that can be studied as a scientific subject"; consequently, he is "careless about himself and other people, including their feelings."[36] Higgins is of this attitude when he encounters Eliza in the street, regarding her merely as a scientific object to be examined as he transcribes and exactly rearticulates her utterances in his own "Universal Alphabet."[37] He remarks to her that "a woman who utters such depressing and disgusting sounds has no right to be anywhere—no right to live." His sense of superiority over her is evident in his assertion that she must "remember that you are a human being with a soul and the divine gift of articulate speech: that your native language is the language of Shakespear [*sic*] and Milton and The Bible; and dont sit there crooning like a bilious pigeon." She responds with an almost indescribable noise (rendered by Shaw as "Ah-ah-ah-ow-ow-ow-oo!"), which is immediately transcribed phonographically by Higgins and read aloud again by him, "reproducing her vowels exactly."[38]

Presently Eliza appears in Higgins's Wimpole Street laboratory, a room arrayed with a variety of phonetic instruments, including a phonograph, singing flames, tuning forks, and a "life-size image of half a human head, shewing in section the vocal organs."[39] Upon her arrival, Higgins remarks to his collaborator Pickering:

> This is rather a bit of luck. I'll shew you how I make records. We'll set her talking: and I'll take it down first in Bell's Visible Speech; then in broad Romic; and then we'll get her on the phonograph so that you can turn her on as often as you like with the written transcript before you.[40]

Higgins's playful words somewhat conceal the multiple purposes of their endeavor. The objective is not simply to transfer Eliza's voice to the phonograph so the men can hear her voice at will—either for Higgins to demonstrate the apparatus to Pickering, or to diagnose what needs "fixing" in Eliza's dialect. Rather, the passage sets up the idea that Eliza must herself become a phonograph, in its wry anthropomorphic figuration of her as a recording to be repeatedly examined.

At one point Mrs Higgins rebukes the two men, and Higgins's response is instructive as to his purpose:

> MRS HIGGINS. You certainly are a pretty pair of babies, playing with your live doll.
>
> HIGGINS. Playing! The hardest job I ever tackled: make no mistake about that, mother. But you have no idea how frightfully interesting it is to take a human being and change her into a quite different human being by creating a new speech for her.[41]

For Eliza does not adopt the tones of another voice as a temporary trick. Her whole person is reshaped, reprogrammed, by the multiple articulatory impressions that her months-long reeducation involves, until she is able to repeat "just like a parrot . . . every possible sort of sound that a human being can make."[42] The process is so closely associated with phonographic recording and replay that Eliza herself seems to merge with the apparatus. The creation of a new speech depends on her physiological re-creation, a transformation that is tracked and exhaustively documented, as Higgins notes: "Every week—every day almost—there is some new change. . . . We keep records of every stage—dozens of gramophone disks and photographs."[43] Her new way of speaking becomes "unconscious," because it has been seared permanently into her body—recalling, perhaps, the etymological origin of the very word "recording" in the notion of learning *by heart.* The process of Eliza's vocal education, in other words, is not simply social or psychological in an abstract sense, but material and physiological. In exceeding Higgins's expectations by completely mastering the art of articulatory transformation, Eliza functions for him as an idealized phonetic machine.

In this sense, *Pygmalion* is very much in keeping with the sculptural theme of the original myth against which the play is quite deliberately placed. Yet, for the reasons that I have already suggested in this essay, the imbrications of phonetic and phonographic culture in the later nineteenth century offer a further important context for the play that helps it exceed its Ovidian frame.[44] John M. Picker has recently shrewdly situated *Pygmalion* in the "trajectory of imaginative representation of the 'female talking machine,'" reading through it a web of connections from Edison's 1880s phonographic talking dolls to E. E. Kellett's 1900 short story "The New Frankenstein," which was abridged and republished the following year in *Pearson's Magazine* as "The Lady Automaton"—a text that Philip Klass has similarly suggested may have been a source for Shaw, with many shared features.[45] In closing, I will build on Picker's argument that Eliza ultimately eludes Higgins's controlling grasp in becoming (as he puts it) "no mere talking machine but an independent speaker who reveals herself to be the upwardly mobile, self-governed voice of Edisonian modernity," by connecting the emergence of universal alphabetics as a practice of vocal education with the claims that Laura Otis has made in suggesting that the play "parodies [fin de siècle] conversion narratives, in which miserable creatures are rehumanized and achieve enlightenment."

For Otis, even as Shaw "challenge[s] the notion that speech distinguishes people from animals," he nevertheless depicts the discomforting process of "transformations [that] cause excruciating pain," perpetrated by a scientific protagonist who gives "little thought to how the [creatures] will live once they have been transformed."[46] In *Pygmalion,* Higgins oscillates between regarding Eliza as a "creature" and as a fellow member of his own species, reminding her that access to the human soul is found only through the physiological and moral enlightenment of language. Eliza does not become creditably human for Higgins until she is physiologically upgraded to an appropriately refined state by her vocal education.

Later in his career, Graham Bell had become irrevocably associated with the oralist movement, whose goal was to assimilate deaf people into society by abolishing the use of signed languages and emphasizing speaking and lipreading in their place.[47] Bell's interventions in the field had an obnoxious (and not, alas, uncommon) eugenic dimension. Jennifer Esmail has shown how deaf people were seen in the nineteenth century "as less than fully human," their linguistic repertoire of visual gestures compared to the rudimentary vocalizations and gesticulations of monkeys and apes. The construction of the deaf as biologically inferior led eugenicist agitators to cast them as a "threat to the 'fitness' of the human race through deaf intermarriage and its potential reproduction of deafness through generations."[48] In such a poisonous culture, it was not clear how oralist demands could possibly be a help to the process of the mutual adaptation and negotiation of hearing people with deaf people. The work that had been notably inaugurated in the theory and practice of Visible Speech led, however circuitously, to a distinctly unpleasant destination. The figuring of speechless subjects as "less than fully human" is implicitly coded in the utopian, normalizing rhetoric that claimed advancement in civilization and global communication would be secured by a universal alphabetics, visibly spoken.

One contemporary reviewer of *Pygmalion*, the radical journalist Henry William Massingham, complained that in place of Ovid's sympathetic protagonist, Shaw's "Pygmalion," Higgins, is "merely a diligent watcher of a test tube."[49] Yet in figuring Eliza as a scientific object that escapes her phonetic laboratory, Shaw's achievement is to sharply ventriloquize Higgins's almost inexpressible discomfort at the rapid social transformation that attended the phonetic in his education of Eliza. The satire becomes most acerbic in Higgins's scientifically obsessed, "careless" self-regard, and painfully mocked in the emotional departure of his charge. The arrogant declaration of his own independence turning to "sudden humility," Higgins confesses, with uncharacteristic sentimentality: "I shall miss you, Eliza. . . . I have learnt something from your idiotic notions . . . I have grown accustomed to your voice and appearance. I like them, rather." Eliza: "Well, you have both of them on your gramophone and in your book of photographs. When you feel lonely without me, you can turn the machine on. It's got no feelings to hurt." In the wake of this wounding riposte, Higgins is left to pathetically implore, "I cant turn your soul on. Leave me those feelings; and you can take away the voice and the face. They are not you."[50] In the end, Eliza is the automaton turned autonomous, the emancipated fugitive of Higgins's overbearing power. One suspects the many students of Visible Speech under the Bells came to enjoy no such enfranchisement from their ordeals.

Conclusion: Physiological Resonance

Long ago, in *Technics and Civilization*, Lewis Mumford diagnosed a profound shift in technological development witnessed by the nineteenth century—a movement away from the enormous machinic scale that distinguished the industrial age, and toward the scale of the human. Since around 1870, he argued, "The organic has become visible again even within the mechanical complex," from which "some of our most characteristic mechanical instruments—the telephone, the phonograph, the motion picture—have grown out of our interest in the human voice and the human eye and our knowledge of their physiology and anatomy."[51] Mumford's assertion that underlying such development is "the effort . . . either to extend the powers of the otherwise unarmed organism, or to manufacture outside the body a set of conditions more favorable toward maintaining its equilibrium and ensuring its survival," has motivated a whole line of inquiry in the study of media, best encapsulated in Marshall McLuhan's axiomatic notion of the "extensions of man."[52] Recently, literary critics and cultural historians alike have begun to attend ever more closely to the ways in which the mechanical and the organic—so often held apart, even defined as outright opposites—might be seen more accurately to have long operated in a system of exchange and feedback.[53] For the technological evolution of mechanical contrivances modeled on organic structures gave rise, in turn, to new and finer understandings of the mechanics of the body.

In this chapter, I have attempted to demonstrate one set of ways in which such understandings were arrived at. As I have shown, the proliferation of inquiry into physiological resonance and auditory vibration in the period immediately before the appearance of the phonograph—a mechanical device that depended for its operation on precisely these principles—converges with a longer lineage of investigation into the physiological basis of vocal movement and memory. Reading these contexts alongside the Bell performances and *Pygmalion*—a stinging satire written in the age of the phonograph but illuminating the sometimes condescending pedagogical culture that preceded it—indicates that the affiliation of the physiological body with a capacity so uniquely cultural as language was often ethically fraught.

Together, these texts and contexts provide some germs of insight into a mode of thinking in which the mind-body was figured as a read-write device years prior to the emergence of the first technical instrument to perform such operations. Alexander Melville Bell's insight and tenacity in pursuing the development of a phonetic alphabet based not on arbitrary symbols but on the movements of the human vocal apparatus prefigures an explosion of physiological inquiry attentive to the performative aspects of the process of memorization. *Pygmalion* offers a collapsing of phonetics into the suggestive notions of resculpting and reshaping. What I have called in this chapter "phonographic physiology" points to one influential nexus of relations from which the recorded and recordable body emerges as a central and enduringly problematic theme of modernity.

Notes

1. See Ira Jean Hirsh, "A Brief History of the Systems Used to Represent English Sounds," *Quarterly Journal of Speech* 29, no. 3 (1943): 334–42; Alexander Melville Bell, *Visible Speech: The Science of Universal Alphabetics, or Self-Interpreting Physiological Letters, for the Writing of All Languages in One Alphabet* (London: Simpkin, Marshall, 1867).

2. Alexander Melville Bell, *The Principles of Elocution; with Exercises and Notations for Pronunciation, Intonation, Emphasis, Gesture and Emotional Expression*, 4th ed. (Salem: Burbank, 1878), xviii.

3. Bell, *Principles of Elocution*, xviii.

4. Alexander Melville Bell, "On Visible Speech: or, a Universal and Self-Interpreting Physiological Alphabet" [1866], Alexander Graham Bell Family Papers, Library of Congress, Washington, DC (hereafter AGB), Box 15, 1.

5. Ibid.

6. A. M. Bell, "Visible Speech," *Werner's Magazine: A Magazine of Expression* 25 (1900): 213.

7. A. M. Bell, lecture notes [n.d.], AGB, Box 12, 12; emphasis added.

8. A. M. Bell, "Visible Speech," 216.

9. "Visible Speech," *Athenæum* 1968 (July 15, 1865): 84.

10. Thomas L. Hankins and Robert J. Silverman, *Instruments and the Imagination* (Princeton, NJ: Princeton University Press, 1995), 178.

11. Alexander John Ellis, *The Alphabet of Nature; or, Contributions towards a More Accurate Analysis and Symbolization of Spoken Sounds; with Some Account of the Principal Phonetical Alphabets Hitherto Proposed* (London: Bagster and Sons, 1845), 25.

12. A. M. Bell, *Visible Speech*, 11.

13. Ibid., 19.

14. Ibid., 22.

15. Alexander John Ellis, "Mr. Melville's 'Visible Speech,'" *Reader* 4, no. 88 (September 3, 1864): 304.

16. Ibid.

17. A. G. Bell, "Prehistoric Telephone Days," *National Geographic Magazine* 41, no. 3 (March 1922): 228.

18. A. M. Bell, "Mr. Melville Bell's 'Visible Speech' No. II," *Reader* 6, no. 136 (August 5, 1865): 155.

19. Robert V. Bruce, *Alexander Graham Bell and the Conquest of Solitude* (Ithaca, NY: Cornell University Press, 1973), 56.

20. A. G. Bell, letter to A. M. Bell, Eliza Symonds Bell, and Carrie Bell, December 1, 1871, AGB, Box 4, 1.

21. Jean-Marie Guyau, "The Origin of the Idea of Time by Jean-Marie Guyau," in *Guyau and the Idea of Time*, ed. John A. Michon, Viviane Pouthas, and Janet L. Jackson (Amsterdam: North-Holland, 1988), 118.

22. Laura Otis, *Organic Memory: History and the Body in the Late Nineteenth and Early Twentieth Centuries* (Lincoln: University of Nebraska Press, 1994), 17.

23. Herbert Spencer, *The Principles of Biology*, vol. 1 (London: Williams and Norgate), 183; see also E. Ray Lankester, "Perigenesis v. Pangenesis: Haeckel's New Theory of Heredity," *Nature* 14, no. 350 (July 13, 1876): 235–38.

24. H. Charlton Bastian, "On the 'Muscular Sense,' and on the Physiology of Thinking," *British Medical Journal*, 1, no. 435 (May 1, 1869): 394.

25. Ibid.

26. Ibid.

27. Theodate L. Smith, "On Muscular Memory," *American Journal of Psychology* 7, no. 4 (July 1896): 454–55; emphasis added.

28. Ibid., 455.

29. Catherine Robson, *Heart Beats: Everyday Life and the Memorized Poem* (Princeton, NJ: Princeton University Press, 2012).

30. See Ashley M. Miller, "Involuntary Metrics and the Physiology of Memory," *Literature Compass* 6, no. 2 (2009): 549–56.

31. Ellis, "Mr. Melville's 'Visible Speech,'" 304.

32. Bernard Shaw, *Pygmalion*, ed. Dan H. Laurence (London: Penguin, 2003).

33. Alexander Bell, *The Bride, a Play, in Five Acts* (London: Cleaver, 1847).

34. See, for instance, Bertrand M. Wainger, "Henry Sweet: Shaw's '*Pygmalion*'," *Studies in Philology* 27, no. 4 (1930): 558–72; Beverley Collins and Inger M. Mees, *The Real Professor Higgins: The Life and Career of Daniel Jones* (Berlin: Mouton de Gruyter, 1999).

35. Shaw, *Pygmalion*, act 1, 17. Shaw tended not to use apostrophes, and the edition used in this essay reflects his nonstandard punctuation.

36. Ibid., act 2, 24.

37. Ibid., act 1, 18.

38. Ibid.

39. Ibid., act 2, 23.

40. Ibid., act 2, 25.

41. Ibid., act 3, 65.

42. Ibid., act 3, 66.

43. Ibid., act 3, 65.

44. Essaka Joshua has demonstrated that Shaw was indeed far more concerned to satirically engage with the "nineteenth-century context" than to construct a direct

theatrical adaptation of the Greek tale. See *Pygmalion and Galatea: The History of a Narrative in English Literature* (Aldershot: Ashgate, 2001), 97–133.

45. John M. Picker, "My Fair Lady Automaton," "Victorian Oral Cultures," special issue of *Zeitschrift für Anglistik und Amerikanistik* 63, no. 1 (2015): 97; Philip Klass, "'The Lady Automaton' by E. E. Kellett: A *Pygmalion* Source?," *Shaw* 2 (1982): 75–100.

46. Laura Otis, "Monkey in the Mirror: The Science of Professor Higgins and Doctor Moreau," "Darwin and Literary Studies," special issue of *Twentieth Century Literature* 55, no. 4 (2009): 485–86.

47. For a longer version of this history, see Jennifer Esmail, *Reading Victorian Deafness: Signs and Sounds in Victorian Literature and Culture* (Athens: Ohio University Press, 2013).

48. Ibid., 104, 134. See also Danielle Coriale's contribution to this volume, which traces the influence of fears about the hereditariness of deafness on the arch-eugenicist Francis Galton, who drew directly on Graham Bell's work in his alarmist review essay "Hereditary Deafness," *Nature* 31 (January 1885): 269–70.

49. Quoted in Essaka Joshua, *Pygmalion and Galatea: The History of a Narrative in English Literature* (Aldershot: Ashgate, 2001), 97.

50. Shaw, *Pygmalion*, act 5, 100.

51. Lewis Mumford, *Technics and Civilization* (1934; Chicago: University of Chicago Press, 2010), 6.

52. Marshall McLuhan, *Understanding Media: The Extensions of Man* (1964; London: Routledge, 2001).

53. For instance, see Tamara Ketabgian, *The Lives of Machines: The Industrial Imaginary in Victorian Literature and Culture* (Ann Arbor: University of Michigan Press, 2011); Jonathan Sterne, *MP3: The Meaning of a Format* (Durham, NC: Duke University Press, 2012).

CHAPTER 7

"So Extraordinary a Bond"

Mesmerism and Sympathetic Identification in Charles Adams's Notting Hill Mystery

Lara Karpenko

In the lead article for the October 1854 issue of the *Zoist*, a relatively short-lived journal dedicated to studying "cerebral physiology and mesmerism," the Reverend R. A. F. Barrett published an account of "A," a "Lady" who was apparently cured by mesmerism after "being twelve years in the horizontal position with extreme suffering."[1] Barrett's account, at once lowbrow entertainment and sober scientific study, casts mesmerism as a legitimate, though extraordinary, medical cure. In the narrative's apparent climax, Barrett details how he managed to keep his starving patient nourished:

> The retching had come on the preceding night as she had predicted. I kept her asleep two hours, and had dinner for her . . . for thirteen days after that she tasted nothing solid. . . . Not only when asleep but when awake also, she seemed to derive real benefit from my eating by her side when she was in mesmeric sleep, and when she awoke could always tell by her own feeling whether I had eaten or not.[2]

As he describes A's ability to receive sustenance from the food that passes his lips, Barrett suggests that mesmerism's effectiveness rests in its ability to form profound, if not titillating, sympathetic bonds. Though Barrett's account may seem fantastic, its focus on sympathy would not have surprised Victorian readers. By the 1850s, discussions of mesmerism consistently intertwined the controversial science with the concept of sympathetic knowledge and identification.

Though the *Zoist* ceased publication in 1856, Barrett's article curiously resurfaced seven years later as a crucial plot device in Charles Adams's *Notting Hill Mystery* (1862–63).[3] Generally overlooked today, *The Notting Hill Mystery* is perhaps best known for Julian Symons's declaration that it (and not Wilkie Collins's *The Moonstone* [1868]) should be classified as the first English detective novel.[4] Though Adams may have beaten Collins to the narrative punch, his novel, a polyphonic page turner, clearly seems inspired by *The Woman in White* (1859) and other early sensation novels. Told from the perspective of Ralph Henderson, a life insurance agent assigned to investigate three untimely deaths and Baron R——'s resulting inheritance, *The Notting Hill Mystery* weaves together diary entries, newspaper clippings, personal letters, various affidavits, and Henderson's increasingly horrified commentary. Ultimately Henderson discovers that the diabolical Baron R callously murdered three people in order to come into his inheritance: his wife, Madame Rosalie; her estranged twin sister, Gertrude Anderton; and Gertrude's husband, William Anderton. A chillingly effective villain, the Baron is remarkably evasive and commits none of the murders directly. Inspired by Barrett's article in the *Zoist* (which Adams reproduces almost in its entirety as part of the novel), the Baron first murders Mrs. Anderton through the sympathetic poisoning of her mesmerized twin sister; he then manipulates Mr. Anderton into committing suicide and finally murders his own wife by throwing Rosalie into a mesmeric sleep and causing her to swallow a deadly poison. Though Henderson is able to piece together the entire crime, his incredulity at the Baron's first murder in particular, and his realization that such a crime, even if possible, could never be prosecuted in court, causes the intrepid insurance agent to abandon the case. In a sharp departure from the sensation novels that defined the midcentury, in *The Notting Hill Mystery*, the criminal seems to get away with murder.

To some extent, however, stating that the Baron "gets away" suggests more closure than the novel actually provides. While the reader never experiences the satisfaction of seeing the Baron brought to justice, the reader also does not see the Baron effectively evade justice. With an ending that almost seems to anticipate "The Lady or the Tiger" in its inconclusivity, *The Notting Hill Mystery* resists the formula that typifies most—if not all—sensation novels. While sensation novels repeatedly end with images of domestic stability—the criminals safely removed and the detective heroes blissfully married—*The Notting Hill Mystery* presents a world in which order seems forever lost.[5] At once conforming to and resisting the formulas that had come to define sensation fiction, *The Notting Hill Mystery* thus defies easy classification.[6] Perhaps because of this, the novel seems to have dropped off our contemporary radar. But when placed within the frame of mesmeric science, Adams's novel showcases a surprising moment in the Victorian history of reading and warrants further critical attention. In some ways, mesmerism's prominence in the novel is unexpected if not anachronistic. By the novel's 1862 publication date, mesmerism, though still a subject of discussion and interest, hardly arrested public attention as it did during the 1840s and 1850s—the decades in which the "mesmeric mania" gripped the public imagination.[7] Despite's mesmerism's fading popularity, Adams casts it as dangerous and deeply disruptive. Indeed, it is the Baron's ability to commit a mesmeric murder that so confounds Henderson and that enables the crimes to remain unprosecuted. Within the novel, then, it is mesmerism and the extraordinary sympathy it represents, and not necessarily the murders themselves, that destabilize the Andertons' domestic world and ultimately the coherence and form of the novel itself.

It is important to keep in mind that for a Victorian audience, the concepts of sympathy, sympathetic identification, and sympathetic bonding were all extraordinarily complex and multifaceted. In her study of sympathetic identification, Rae Greiner aptly argues that for many eighteenth- and nineteenth-century critics, sympathy was a "form of *thinking* geared towards others" and goes on to suggest that "[emotional] feeling played no . . . necessary part" in the sympathetic experience; in a similar vein, Rachel Ablow suggests that she is "less interested in sympathy as a feeling . . . than in sympathy as *psychic* structure through which the subject is produced, consolidated, or redefined."[8] Though Greiner and Ablow certainly have different points of focus, both notably seek to disentangle sympathy from feeling. By contrast, while I acknowledge that sympathy and emotional feeling were not necessarily connected during the nineteenth century, I suggest that sympathy was often connected with physical feeling. Sympathy was not only a "psychic phenomen[on]," as Ablow suggests, but a physiological phenomenon—one that could erase physical boundaries, destroy bodily integrity, and infect populations.[9] Sympathy, in other words, is a physical experience of connection and community. In defining sympathy in such a way, I intentionally follow

Adams's own use of the term. For instance, near the opening of *The Notting Hill Mystery*, one of the many narrators remarks on the "wonderful sympathy that existed between the twins" (Mrs. Anderton and Madam Rosalie as children). "This sympathy," the narrator states, "seems even more physical than mental . . . every little ailment that affects the one is immediately felt by the other."[10] Building off of Victorian understandings of mesmerism, Adams conceives of the sympathetic experience as dangerously grounded in the body, and as frighteningly communal.

By focusing on the physiological dimensions of the sympathetic relationship, I connect the "mesmeric mania" of the 1840s with the rise of sensation fiction in the 1860s and suggest that sensation fiction promised a sanitized version of the enticing sympathy that mesmerism once offered to its participants.[11] I then turn to *The Notting Hill Mystery* in order to provide a case study in which an author deliberately critiques this Victorian tendency to transform texts into physiological experience. I argue that Adams makes fictional use of the fading strange science in order to warn against the seductive powers of sympathetic identification and, by extension, against the embodied pleasure and sympathetically engaged reading style that sensation fiction explicitly encouraged. Adams's strident critique of sympathetic identification ultimately reveals that for a Victorian audience, reading itself often constituted a strange form of physiological inquiry.

Mesmerism, Sympathetic Knowledge, and the Reader of Sensation Fiction

Before moving on to discuss Adams's critique of sympathetically engaged reading, I begin with an overview of the mesmeric construction of sympathetic identification. Though practitioners and critics of mesmerism agreed on very little (a point beautifully documented by Alison Winter), both camps acknowledged that mesmerism facilitated intense physiological bonds.[12] Certainly the very structure of the mesmeric séance, which generally featured a male mesmerist and a female subject, assumed that sympathetic (i.e., physical) bonding was inseparable from the experience. In order to describe the dynamic between mesmerist and mesmerized, I turn to a report from the *Lancet* that describes one of the most well-known mesmeric demonstrations of the era—Dr. John Elliotson's first public experiment on Elizabeth O'Key. Though Elliotson's experiments on the O'Key sisters were eventually discredited by the scientific community in general and (vociferously) by the *Lancet* in particular, Elliotson, whom Fred Kaplan decrees "most responsible for the spread of mesmerism in England," nonetheless helped establish the protocols for the mesmeric séance:[13]

> She was here put to sleep (this was always done by a pass of the hand) . . . the Doctor drew his hand, pointed towards hers, upwards and outwards in the air. In a few seconds her hand and arm began to move up in the same direction. While ceasing, for a short time, in order to talk to someone near, he produced a motion with his fingers, which those of the girl immediately imitated. "See," said the Doctor, "my fingers were moved involuntarily; I did not mean to influences hers."[14]

It is this "pass"—or the slow gliding of the hands over, but not necessarily touching, the body of the subject—that became the hallmark of the mesmeric experience. A gesture at once intimate and public, the pass signaled that bodily boundaries were permeable and that doctor and patient had enticingly become one. As Elliotson demonstrates mesmerism's validity by encouraging O'Key's movements to mirror his own and as the *Lancet* reporter seems particularly impressed when O'Key mirrors even Elliotson's unconscious movements, it becomes clear that the very goal of the mesmeric séance was embodied experience and sympathetic communion.

In perhaps an even more direct conflation of mesmeric practice with sympathetic bonding, Thomas Buckland offers the following piece of advice in his *Handbook of Mesmerism* (1850): "In order that one individual may [mesmerize] . . . another," he writes, "there must exist between them a moral and physical sympathy."[15] Explicitly suggesting that mesmerism requires a *physically* sympathetic connection, Buckland's manual, a simple "how-to" guide aimed at the mesmeric layman, suggests just how necessary this conception of the permeable body was to the mesmeric experience. Notably, it was not only mesmerists who encouraged sympathetic bonding. Mesmerized subjects also defined such bonding as one of the preconditions and as one of the benefits of the mesmeric séance. For instance, in her controversial *Letters on Mesmerism* (1845), Harriet Martineau, reflecting on her own experiences as a mesmerized subject, insists that mesmerism forges bonds so intense that the "sympathy induced by two or more persons resemble[s] no other relation known."[16] In a time when, as Ablow points out, sympathetic bonding "came to function . . . as a pleasurable characteristic of the domestic sphere" and to some extent defined the companionate marriage, Martineau casts mesmeric sympathy, a sympathy that was realized in the body, as capable of supplanting even domestic or marital bonds.[17]

Perhaps because Martineau understood mesmerism sympathetically, or perhaps because she foresaw that her account would inspire critique, she actually changed mesmerists twice before 1845: Martineau began her sessions with the itinerant male mesmerist Spencer Hall, switched to her personal maid, and finally chose "a lady, the widow of a clergyman . . . [with] high qualities of mind and heart" to perform her regular mesmeric treatments.[18] Despite the care that Martineau took in selecting her mesmerist, her account inspired widespread critique—much of it centering on the very issue of sympathetic melding. In a particularly scathing review, a critic who only called himself "Veritas" suggests that Martineau, one of the most celebrated public intellectuals of her decade, is a dupe whose "judgment is perverted [and] mental faculties obscured," and, not surprisingly, he accuses her mesmerist of obscuring Martineau's mental faculties.[19] "Miss M.," Veritas suggests, "appears to be imposed upon, and is unconsciously lending herself to impose upon others."[20] Though Veritas proudly defines himself as an opponent of mesmerism, he nonetheless casts mesmerism as dangerously sympathetic; to some extent, Veritas seems to suggest that mesmerism has sympathetic potential *because* it is invalid. As the very term "impose" suggests, embedded in Veritas's critique is the fear that the charlatanical mesmerist's desires will multiply and ripple throughout the British public.[21]

Though Martineau's account was indeed controversial, the enthusiasm for mesmerism continued throughout the 1840s, and by 1851, John Bennett anxiously proclaimed that a "mesmeric mania" had overtaken Great Britain and casts mesmerism as both a physiological experience and as frighteningly sympathetic:

> I have been told that in some educational establishments, girls and boys throw themselves in to states of trance and ecstasy, or show their fixed eyeballs and rigid limbs, for the amusement of their companions. Sensitive ladies do not object to indulge in the emotions so occasioned, and to exhibit themselves in a like way for the entertainment of evening parties. . . . The disorder has not been confined to Edinburgh. . . . [Dr. Darling] has produced the greatest excitement in his course toward London, where according to the papers, there are at present repeated the same public scenes, and the same phenomena,

> as were produced among us . . . so that, I think, we are warranted in calling it—"The Mesmeric Mania of 1851."[22]

Embedded in Bennett's critique of mesmerism is his awareness that its adherents seem to experience the phenomena in their bodies. Indeed, as he describes the "girls and boys" who display their ecstatic bodies and the "sensitive ladies" who forgo modesty in their eagerness to exhibit themselves, Bennett seems to define mesmerism as not just embodied but as excessively embodied. Comparing mesmerism to a contagion, Bennett acknowledges the sympathetic powers of mesmerism; within Bennett's description, mesmerism almost seems to leap from person to person, from doctor to patient from city to city and even from country to country. In some ways, however, Bennett's description of the "Mesmeric Mania of 1851" signifies the end—rather than the height—of serious interest in mesmerism. The very public critiques of Martineau did perhaps suggest that the science, with its promises of sympathetic identification and of an intensely embodied experience, was not fit for the middle and upper classes. Mesmerism's fading appeal was revealed most prominently by Elliotson's decision to cease publication of the *Zoist* in January 1856. Though Elliotson claimed that he was ending publication because "the object for which *The Zoist* was undertaken is attained" and though some interest in mesmerism certainly continued throughout the nineteenth century and continues even into the twenty-first century, it was clear that mesmerism was no longer the subject of fervid public fascination.[23] However, by 1859, with Wilkie Collins's landmark publication of the *Woman in White*, mesmeric language resurfaced once again in reviews and responses to the sensation novel. The interest in mesmerism, as Winter and Garrison also relate, did not fade so much as it become rerouted. Mesmerism became textual.[24]

Sensation fiction thus took up the "strange" cultural work once performed by mesmerism by providing the intriguing possibility of a stimulated body. The sensational author becomes a sort of "textual mesmerist"—diffusing his (and less frequently her) desires in order to bind readers to the novel's characters and to one another. In a facetious though telling example, *Punch* announced in 1863 that a new journal would begin publication: *The Sensation Times and Chronicle of Excitement.* Beginning by suggesting that sensation novels had become "one of the necessities of the age," the "advertisement" wryly caricatures the objectives of sensation fiction. "This Journal," spoofs *Punch,* "will be devoted chiefly to the following objects: namely, Harrowing the Mind, Making the Flesh Creep, Causing the Hair to Stand on End, Giving Shock to the Nervous System, Destroying Conventional Moralities, and generally Unfitting the Public for the Prosaic Avocations of Life."[25] As the anonymous author jokingly points to the nervous system, the flesh, and the hair, he both echoes the public discourse surrounding mesmerism and describes the reading experience itself as requiring physiological diagnosis. Whereas previous subgenres of fiction certainly engaged the body (sentimental fiction elicited tears for instance), that readerly experience was, as Winter points out, "mediated by . . . judgment and imagination"; by contrast, with sensation fiction, as Winter argues, "The route from page to nerve was direct."[26]

Perhaps no other Victorian literary form addresses the physiological aspects of textual mesmerism more directly than literary reviews, and I close this section by briefly examining Margaret Oliphant's oft-cited essay "Sensation Fiction." While both Garrison and Winter point out that critical reviews of early sensation novels generally feature language that mirrors mesmeric discourse, neither focuses directly on how these reviews construct and participate in the construction of physiological sympathy. Taking Oliphant's review as representative of Victorian literary reviews more largely, I build off of Nicholas Dames's astute observation that critics consciously sought to reveal the "effect of reading."[27] Appearing in 1862, almost three years after the *Woman in White* began its serial run, Oliphant's piece functions more as a retrospective on sensation fiction's rapid development than as a review. Beginning with Collins's *Woman in White* and readily admitting that "we need not discuss over again so familiar a tale," Oliphant indicates that her purpose is to alert readers not to a new novel but to a new reading practice.[28]

Extensively quoting from the *Woman in White*, Oliphant goes on to explain not *what* the novel means but *how* the novel means.[29] Though readers today might find the Victorian convention of the prolonged quotation tedious, Dames suggests that the practice was integral to nineteenth-century critical protocols and was meant to provide "a reading experience in miniature . . . [to invite] the review's reader to partake in a reading experience with the critic."[30] In particular, the prolonged excerpt allows Oliphant to record the physiological effects of the novel. After quoting nearly the entire passage of Walter Hartright's initial encounter with the eponymous "woman in white" (arguably the scene that that inaugurated the entire subgenre of sensation fiction), Oliphant immediately turns her attention to readerly reaction. "Few readers," she suggests, "will be able to resist the mysterious thrill of this sudden touch.

The sensation is distinct and indisputable. The silent woman lays her hand upon our shoulder as well as upon that of Mr. Walter Hartright."[31] By explicitly conflating the reader with Hartright, Oliphant suggests that one of the thrills of the novel is that of sympathetic identification; within Oliphant's formulation, the reader not only identifies with Walter Hartright, but in some senses becomes him, enticingly and perhaps troublingly erasing the boundary between self and character. Further, Oliphant notably uses the plural form "our" to speak of the readerly experience—a word that at once suggests plurality and unity. The *Woman in White*'s many readers, Oliphant thus seems to hint, experience the novel in only one uniform way. Just as mesmerizer and mesmerized excitingly meld into one enervated entity, the community of sensation readers fuse and blend into one another, feeling one another's sensations, and speaking, as Oliphant does, for one another's reading experiences. While Oliphant may warn against "a kind of literature which must . . . more or less, make the criminal its hero," she also acknowledges that a sympathetic reaction to the texts in which the "reader's nerves are affected like the hero's" is the only "normal" reaction—that is, the reaction experienced by herself and the audience, a reaction that "few readers can resist."[32] Oliphant's insistence on the universality of her physiological reading experience thus obliges readers either to join her or to acknowledge their own physiological differences. In a sense, Oliphant casts the experience of reading sensation fiction as a medical diagnostic instrument: reading becomes a method by which readers can judge whether they have a normal body and whether their physiological reaction is reflected in the population at large.

Though Oliphant does indicate some level of discomfort with the genre, reading sensation fiction was ultimately an acceptable activity, as middle- and upper-class Victorian readers repeatedly recounted their intense reactions to sensation novels. Perhaps one of the *Woman in White*'s greatest fans, Edward Fitzgerald (who even considered naming his boat "Marian Halcombe") suggested that the novel "exerts a sort of magnetism in drawing me toward the corner of a dark Cupboard, or Closet, in which . . . she lies."[33] Casting the novel as a mesmerist who exerts a mysterious influence, Fitzgerald's letter, like Oliphant's review, focuses on the affective dimensions of the novel. While Martineau's published account of inhabiting a mesmerized body was viciously punished, readers like Fitzgerald delightedly, and almost obsessively, detailed their profound, if not intimate, bodily experiences in reaction to sensation novels. Indeed, as respectable journals like Dickens's *All the Year Round* published them, reputable men like Edward Fitzgerald read them, and middle-class families regularly bought them, it appears that sensation novels, despite the seemingly salacious pleasures they offered, were embraced by a large segment of the Victorian reading population. Responding to this popularity of sympathetic and embodied reading practices, Adams draws on the strange science of mesmerism in order to recall and resist sensation novels. In a decade when readers eagerly sought out fictional experiences that would inspire them to mimic the nervous, excited, and panting bodies represented in the pages of a novel, mesmeric science and the backlash it inspired provided Adams with a convenient frame to cast sympathy as a physically destructive and invasive force.

The Notting Hill Mystery, *the Mesmerizing Villain, and the Dangers of Sympathy*

Originally serialized in eight parts in *Once a Week* (published as a single volume in 1863) and featuring illustrations by George du Maurier, *The Notting Hill Mystery* appears to contain all the elements of a sensation novel: a peaceful middle-class home destroyed by crime, mistaken identity, poison, and an intricate plot. Most particularly, in what seems to recall Collins's depiction of Walter Hartright and Frederick Fairlie, or Ellen Wood's depiction of Isabel Vane, Adams highlights the physical and mental weakness of his victim heroes, Gertrude and William Anderton. For instance, Henderson, who serves as the novel's primary narrator, describes "William's constitutional nervousness, mental as well as physical" (38), while various characters refer to Gertrude's "delicate" constitution (34, 37, 41, 264). By contrast, the villainous Baron, though "a little stout squab man" (67), is undeniably debonair and magnetic—almost effortlessly fooling those he comes in to contact with. "Everybody liked him," comments Mrs. Brown, his ever-admiring landlady. "He was so good natured" (142). Though Gertrude initially worries that the Baron "would [not] have much compunction in killing anyone who offended him, or who stood in his way," she quickly becomes fascinated by his "wonderful green eyes" and agrees to his mesmeric trials (69). Though *The Notting Hill Mystery* predated *The Moonstone*, Adams notably published his novel after *The Woman in White* and *East Lynne* had mesmerized the Victorian public and primed an entire readership to associate enervated characters and charming villains with the genre of sensation fiction. Certainly, as Henderson begins his account with the proclamation that he is "laying before [the reader] . . . extraordinary revelations" (1) and promises to reveal an "unfathomable mystery" (10), the novel seems to be almost in lockstep with other sensation narratives.

In its representation of sympathetic community however, *The Notting Hill Mystery* differs sharply from the trajectory of other sensation novels. While sympathetic communion generally facilitates detection (i.e., Marian Halcombe's and Water Hartright's sympathetic understanding of Laura Fairlie propels them to uncover Fosco's crimes), in *The Notting Hill Mystery*, almost all forms of connection or communal bonding facilitate the Baron's murderous plans. From large-scale popular events, like the Crystal Palace exhibition, that force the Andertons to interact with the Baron, to Rosalie and Gertrude's intimate, "extraordinary," and ultimately deadly twinship bond (31), Adams links all communal experiences with tragic consequences. Even the simple act of human touch, as shown most particularly in du Maurier's dark and moody illustrations for the novel, only seems to bring pain. In perhaps the most arresting of these illustrations, du Maurier portrays one of the Baron's initial attempts to fuse Gertrude, Rosalie, and himself during a séance. At the foreground of the picture lie the mesmeric trio, while Mr. Anderton and his friend, Frederick Morton, observe in the background. Ominously standing over his wife, the Baron makes a mesmeric pass with his right hand while firmly holding Rosalie with his left, while Rosalie, in turn, holds Gertrude. As they lie back passively, Gertrude and Rosalie clearly accept the Baron's touch (even if only—as in Gertrude's case—by proxy) and thus signal their willingness to accept his mental and physiological imprint. In contrast to the vibrantly fused trio in the foreground, William and Frederick appear completely passive and almost indistinguishable from the background, signaling that the husband-wife bond has been symbolically broken. The Baron, the illustration suggests, uses touch to create an alternate, parodic family structure. Hearkening back to the iconic moment when Anne Catherick laid her ghostly hand on Walter's shoulder, and recalling Martineau's suggestion that mesmeric bonds could supplant domestic ones, sensation, in its most literal form, dissolves mental, physical, and familial integrity.

Of course, as du Maurier's illustration indicates, mesmerism represents the form of bonding that is most dangerous in the novel. After the naive Andertons begin their mesmeric practices, the novel quickly begins to splinter in terms of both narrative coherence and structure. In a significant departure from the plot conventions of sensation fiction, all the characters who could be termed the novel's heroes (Ger-

Fig. 7.1. Illustration by George du Maurier, from *The Notting Hill Mystery* by Charles Warren Adams, *Once a Week: An Illustrated Miscellany of Literature, Art, Science, and Popular Information* 7 (1862): 645. (Courtesy of University of Michigan and Carroll Libraries.)

trude Anderton, William Anderton, Madame Rosalie) are murdered by the Baron and, notably, the two female characters are murdered through mesmeric means. Peculiar susceptible to the Baron's powers, Gertrude belies her extreme capacity for physical sympathy with his very first mesmeric experiment. "He certainly is powerful," she writes in her diary, "for he had scarcely made a pass over me before I felt a glow though my whole frame."[34] Though Gertrude seems anxious about the Baron's power, she quickly responds to the pass and allows herself to sympathetically meld with the Baron—unwittingly (and self-destructively) yielding to his desires and accomplishing his pecuniary goals. It is worth noting that early in the novel, the Baron, though aware of Gertrude's inheritance and that Gertrude and Rosalie are estranged identical twins (a fact that curiously eludes every other character in the novel), is unable to figure out how to secure Gertrude's money for himself. It is only after William reads aloud the *Zoist* article, discussed earlier, that the Baron hits on the scheme of mesmerically linking the twins and of sympathetically poisoning Gertrude through the actual poisoning of her physically stronger twin sister (52–53). It is, then, the presence of the *Zoist* in the Anderton home and their willingness to consult a mesmerist at all that facilitates not only Gertrude's tragic death but the Baron's ability to evade detection.

While untimely deaths may have been a feature of the genre (e.g., Anne Catherick in *The Woman in White*), rarely (if ever) do the main characters suffer the prolonged and horrific deaths experienced by Gertrude, William, or Rosalie. Perhaps most haunting is Gertrude's slow and painful poisoning, which Adams depicts in scenes that last for pages. In the first of such scenes, the character of Dr. Watson describes the patient:

> Mrs. Anderton was on the couch in her dressing-room, partially undressed. . . . Almost immediately on my arrival this disturbance recommenced, though there appeared to be now hardly anything left in the stomach. The sickness continued with unabated violence for more than an hour after the stomach had been completely emptied, and was accompanied with other internal derangement and severe cramps both in the stomach and extremities. (82)

As Adams describes Gertrude's partially unclothed body, her violent vomiting, eventually even referring to her "diarrhea" (172), his account is considerably more graphic than even Collins's and Wood's most sensational moments and seems gratuitous in its sharp departures from standards of conventional Victorian taste. Appearing in stark contrast to the typical communal Victorian death- or sickbed scene, Adams's portrayal of Mrs. Anderton's death instead emphasizes fragmentation, discontinuity, and a world in which the domestic order is forever lost.

In another departure from the sensation genre, the three heroes "stay dead" (there is no Laura Fairlie–style return from the brink of the grave here) and their untimely deaths are never avenged by the justice system. Instead of ending with a comforting declarative sentence assuring the readers that stability has resumed, the novel ends with a jarring interrogative. "Are crimes thus committed susceptible of proof," asks Henderson, "or, even if proved, are they of a kind for which the criminal can be brought to punishment?" (284). As Henderson's final question echoes hollowly at the novel's close, the novel's intricately woven plot does not conclude as much as stop. There is no afterword, no indication of what happens to the Baron, and no hint that the Andertons are mourned or missed by anyone. Instead, all plot points and characters seem to evaporate as Henderson hastily undermines the outcomes of his own investigation.

To some extent, the structural irregularities of the novel are even more surprising than its content departures. For the first six parts, the novel is characterized by rapid—almost staccato—shifts in character perspective: twenty-four discrete characters contribute independent narratives in only about two hundred pages. Because even the main characters pass through the narrative so quickly, the sort of readerly identification that Oliphant highlights in her review of *The Woman in White* is rendered almost impossible. Further, by the novel's seventh part, the narrative becomes so splintered as to be nearly experimental. Under the guise of building his case, Henderson increasingly recedes as narrator as he provides ever-more disconnected pieces of textual evidence. It is with Henderson's tenth piece of evidence—a short fragment of a letter found in the French-speaking Madame Rosalie's room—that any narratorial voice seems completely absent. Comprising over three pages of the text, this fictional letter is displayed in four distinct ways: as a pictorial representation of the letter fragment, as an incomplete French version, as a reconstructed French version, and finally as a complete English translation (239–42). Of these four versions, both the pictorial representation and the fragmented French version are incomprehensible to any reader, while the complete French translation would be comprehensible only to a portion of Adam's readers; so, to some extent, Adams offers three pages of text that are undecipherable. Though there is perhaps a puzzle-like element in this section, the fragmentation, repetition, and incomprehensibility of the text forces the reader out of the narrative. Adams never provides a structure to unify, connect, or comment upon these pieces of evidence; instead, in a move that serves to further sever readerly identification and resist mesmeric absorption, the novel moves abruptly to the next piece of evidence.

It is this next piece of evidence—the reproduction of almost the entirety of Barrett's *Zoist* article—that is most jarring to the narrative (242–44). Notably, as varied and wide-ranging as the novel's polyphonic accounts may be, they are all the fictional creations of Adams himself. By contrast, the extensive quotation from the *Zoist* remains the only section of the novel penned by an outside party. Nearly three pages in length, the quotation disrupts not only the narrative structure of the novel but supplants Adams's authorial voice. Though the inclusion of the quotation is loosely justified by Henderson as forming "the concluding portion of the evidence" against the Baron, Henderson also admits that "the bearing of the [*Zoist* document] on the case will perhaps be less clear" (210). While the lengthy extract may clinch the Baron's guilt, it appears at a point at which it is clear that the Baron committed the murders and so adds little to the narrative progression of the novel. As with the letter fragments just before, the extract abruptly ends the chapter and garners no further commentary from Henderson. By this point in the novel, Henderson and all the characters (even the villainous Baron) seem to fade from the narrative altogether, leaving only this article that is external to the novel. In this sense the text's narrative structure mirrors the plot: just as the Andertons' lives never recover after mesmerism enters their household, the narrative splinters after mesmerism enters the text.

Taken together, Rosalie's letter and the *Zoist* extract form a block of text that is disorienting—an effect that is all the more profound because they compose the entirety of two adjacent chapters. Rather than brushing these textual anomalies off as poor writing, I suggest that they are purposeful and serve to reorient Victorian reading practices. Sensation fiction, as Oliphant notes with some anxiety, relies on a rapid, almost frenzied reading style and on enervated, compelling characters that demand a sympathetic response from the reader: the ideal sensation reader, as Edward Fitzgerald indicates, was fully absorbed. By contrast, Adams offers a novel that forces breaks in the reading pace, prevents readers from actively identifying with any one character, and ultimately resists this sort of deep absorption. A novel that appeals to the reasoning mind rather than the sympathizing body, *The Notting Hill Mystery* intervenes in dominant Victorian understandings of reading and anticipates what Dames suggests is a post-Jamesian theory of the novel—or the notion that the novel should function "as an engine for the production of knowledge" rather than as a "machine for the production of affect."[35] Indeed, as Paul Collins comments in his *New York Times* book review, *The Notting Hill Mystery* "is both utterly of its time and utterly ahead of it."[36]

Certainly, contemporary reviews of the novel, though generally positive, attest to the difficulty Victorian readers had in connecting or responding to Adams's anomalous text. For instance, a critic for the *London Review* compliments Adams for offering a "carefully-prepared chaos" but then critiques the novel for failing to provide "the magnetic influence of life-like character." "We are," the critic goes on to complain, "not interested in the victims of this foreign scamp" (178), in effect critiquing the novel for preventing readerly sympathetic identification.[37] In a more positive review for the *London Standard*, the critic opens with a somewhat unexpected opening line, "The best fish sauce ever invented is said to be a combination of all other known condiments."[38] Though this analogy is surely meant as a compliment attesting to Adams's "consummate skill" in crafting a complex story from disparate elements, the very strangeness of the comparison suggests the extent to which the novel remained fundamentally incomprehensible to Victorian readers.[39] While comparisons between novels and food were not unheard of (Wilkie Collins's readers frequently would describe the experience of devouring his novels),[40] the *Standard* reviewer's comparison of *The Noting Hill Mystery* to fish sauce nonetheless remain atypical; unlike the passionate conflation of reading and eating in other reviews, this particular comparison reveals nothing about the novel's readability but only about its undefinability. "It is not a mere sensation novel," declares the critic, "nor an extract from the 'Newgate Calendar' veneered with a . . . Braddon varnish, not a Wilkie Collins romance . . . nor a story founded on some overpowering and inscrutably mysterious plot."[41] Though the reviewer ultimately suggests that *The Notting Hill Mystery* "has an aroma of all these different ingredients" and thus seems to hint that the novel is both like and unlike the sensation novels the reviewer subtly mentions, he or she is, as these apophatic definitions suggest, almost spectacularly unable to define or explain the novel in any comprehensible way. Significantly, in a departure from Victorian convention, neither of the above reviewers provided extended quotations from Adams. While the critic for the *London Review* provides no quotations at all, the critic from the *Standard* only quotes Barrett's *Zoist* article. The rather surprising choice to quote from something that the novelist did not write (and the *Standard*'s critic is quite clear that the extract is from the *Zoist*) suggests not only that the *Zoist* extract is jarring, but that Adams's novel was strange to Victorian readers. If the prolonged excerpt signaled how a text "feels," as Dames argues, the lack of quotation may perhaps indicate that critics (and even readers more largely) did not know how to feel when encountering *The Notting Hill Mystery.*

In creating a text that contains the expected elements of sensation novels and yet also a text in which touch, popular spectacles, genetic kinship, and (most powerfully) mesmerism all signal a dangerous loss of volition, Adams thus comments on the dangers of sensation reading. Focusing particularly on the dangers of sympathetic, communal identification, Adams simultaneously recalls and critiques sensation fiction's biggest lure. If, as Dames so persuasively argues, all reading at midcentury was understood to be embodied (and if reading sensation fiction was *particularly* embodied), *The Notting Hill Mystery* seems to acknowledge this Victorian physiological understanding of reading even as Adams tries to reinvent reading practices.[42] Today mesmerism has been largely relegated to the category of the pseudoscientific, or sometimes as a mirror into the wackiness that often characterized the Victorian era; by contrast, my study underscores the interdependencies between the sciences and the arts during this period, suggesting that many of these strange sciences had cultural impacts far beyond the brief moments in which they were embraced as legitimate.

Notes

1. R. A. F. Barrett, "Mesmeric Cure of a Lady Who Had Been Twelve Years in the Horizontal Position with Extreme Suffering," *Zoist: A Journal of Cerebral Physiology & Mesmerism, and Their Applications to Human Welfare* 12, no. 47 (1854): 213. Hereafter referenced only as *Zoist.*

2. Ibid., 235–37.

3. Charles Warren Adams, *The Notting Hill Mystery*, ed. Mike Ashley (London: British Library, 2012). The text was originally published under the pseudonym "Charles Felix," and it wasn't until 2011 that Adams was identified as the author. Mike Ashley, introduction to *The Notting Hill Mystery*, xiv.

4. Julian Symons, *Moral Consequences: A History from the Detective Story to the Crime Novel* (New York: Harper & Row, 1972), 47.

5. For instance, *The Woman in White* closes with Laura and Walter's marriage and Marian's firm relegation to the status of the "angel of [the couple's] lives." Wilkie Collins, *The Woman in White*, ed. John Sutherland (Oxford: Oxford University Press, 2008), 643. Similarly, Mrs. Henry Wood's *East Lynne* (1861) depicts the tender embraces of Archibald and Barbara as they both swear to forget the adulterous Isabel Vane. Ellen Wood (Mrs. Henry Wood), *East Lynne*, ed. Elizabeth Jay (Oxford: Oxford World's Classics, 2008).

6. This is not to suggest that sensation fiction is a necessarily conservative genre. As I have argued elsewhere, sensation fiction often reimagined traditional Victorian subjectivities. Lara Karpenko, "'A Nasty Thumping at the Top of Your Head': Muscularity, Masculinity, and Physical Reading in *The Moonstone*," *Victorian Review* 38, no. 1 (2012): 133–54.

7. Alison Winter, *Mesmerized: Powers of Mind in Victorian Britain* (Chicago: University of Chicago Press, 1998), 284.

8. Rae Greiner, *Sympathetic Realism in Nineteenth-Century British Fiction* (Baltimore: John Hopkins University Press, 2012), 1, 2–3; emphasis added; Rachel Ablow, *The Marriage of Mind: Reading Sympathy in the Victorian Marriage Plot* (Stanford: Stanford University Press, 2007), 2; emphasis added. Of course Greiner and Ablow are not the only two critics to discuss sympathy. For more on the nineteenth-century construction of sympathy, please see Anne Cvetkovich, *Mixed Feelings: Feminism, Mass Culture, and Victorian Sensationalism* (New Brunswick, NJ: Rutgers University Press, 1992); Audrey Jaffe, *Scenes of Sympathy: Identity and Representation in Victorian Fiction* (Ithaca, NY: Cornell University Press, 2000); Jim Endersby, "Sympathetic Science: Charles Darwin, Joseph Hooker, and the Passions of Victorian Naturalists," *Victorian Studies* 51, no. 2

(2009): 299–320; Anna L. Kornbluh, "The Economic Problem of Sympathy: Parabasis, Interest, and Realist Form in *Middlemarch*," *ELH* 77, no. 4 (2010): 941–67; Christopher Lane, "Bulwer's Misanthropes and the Limits of Victorian Sympathy," *Victorian Studies* 44, no. 4 (2002): 597–624. Please note that the preceding list is not exhaustive. In my discussion, I also draw upon the insights of Cvetkovich, Winter, and Nicholas Dames in suggesting that sympathetic reading was a physically engaging practice. Nicholas Dames, *The Physiology of the Novel: Reading, Neural Science, and the Form of Victorian Fiction* (Oxford: Oxford University Press, 2007), 11.

9. Ablow, *Marriage of Minds,* 7.

10. Adams, *The Notting Hill Mystery,* 26. Page numbers for subsequent citations will be given parenthetically in the text.

11. By suggesting that sensation fiction functioned as an extension of the mesmerism movement, I extend the work of Winter, who provocatively ends her landmark study of mesmerism by turning to *The Woman in White.* "The link of sensation fiction," suggests Winter, "between reader and character recalled the mesmeric 'community of sensation' that removed boundaries between people." Winter, *Mesmerized,* 326. In a similar vein, Laurie Garrison more recently argued that "like sensation novels, mesmerism . . . [was] centrally concerned with senses, bodies, and pleasures, which were usually theorized as elements of the trance state." Laurie Garrison, *Science, Sexuality and Sensation Novels: Pleasures of the Senses* (London: Palgrave Macmillan, 2011), xiv. Both Garrison and Winter aptly suggest that mesmerism and sensation fiction allowed (if not forced) its participants to imagine a world in which interpersonal boundaries are erased. Though I do not dispute Winter's or Garrison's claims in this piece, neither critic examines any novels that disputed or resisted sensation fiction's connection to the mesmeric séance.

12. Winter, *Mesmerized.*

13. Fred Kaplan, "'The Mesmeric Mania': The Early Victorians and Animal Magnetism," *Journal of the History of Ideas* 35, no. 4 (1974): 696.

14. "University College Hospital: Animal Magnetism," *Lancet* 2 (1837–38): 284–85.

15. Thomas Buckland, *The Handbook of Mesmerism: For the Guidance and Instruction of All Persons Who Desire to Practise Mesmerism for the Cure of Diseases, and to Alleviate the Sufferings of Their Fellow Creatures* (London: H. Bailliere, 1850), 9.

16. Harriet Martineau, *Letters on Mesmerism* (London: Edward Moxon, 1845), 39.

17. Ablow, *Marriage of Minds,* 4.

18. Martineau, *Letters on Mesmerism,* 12.

19. Veritas, "Miss Martineau and Mesmerism," *London Medical Gazette* 1 (1837–38): 426.

20. Ibid.

21. As Winter so well documents, Martineau's mesmeric experiments ultimately incurred intense ridicule; most humiliating for Martineau was her brother-in-law Thomas Greenhow's publicly circulated report to the Royal College of Surgeons, "The Case of Miss H—— M——." Winter, *Mesmerized,* 213–30. Asserting that Martineau's mesmeric cure "acted[ed] though the imagination and the will," Greenhow graphically describes Martineau's uterine tumor, effectively suggesting that only the medical community, not Martineau or her mesmerist, can claim knowledge of Martineau's body. Though his study may represent what Frawley describes as a typical Victorian "example of the 'case study' model of medical diagnosis," Greenhow, with his laughable nod to anonymity (any reader would have known that the initials H.M.

referred to Harriet Martineau) and his gratuitous description of her various gynecological exams seems to maliciously punish Martineau for her frank embrace of mesmerism. T. M. Greenhow, "Medical Report of The Case of Miss H—— M——," in *Life in the Sick-Room* by Harriet Martineau, ed. Maria H. Frawley (Peterborough, ON: Broadview Press, 2003), 188, 187.

22. John Hughs Bennett, *The Mesmeric Mania of 1851, with a Physiological Explanation of the Phenomena Produced* (Edinburgh: Sutherland and Knox, 1851), 5–6.

23. John Elliotson, "Conclusion of the *Zoist*," *Zoist* 13, no. 49 (1856): 411.

24. Laurie Garrison, *Science, Sexuality and Sensation Novels: Pleasures of the Senses* (London: Palgrave Macmillan, 2011).

25. "Prospectus of a New Journal," *Punch or the London Charivari* 44 (1863): 193.

26. Winter, *Mesmerized,* 324.

27. Nicholas Dames, "On Not Close Reading: The Prolonged Excerpt as Victorian Critical Protocol," in *The Feeling of Reading,* ed. Rachel Ablow (Ann Arbor: University of Michigan Press, 2010), 17.

28. Margaret Oliphant, *Blackwood's Edinburgh Magazine* 102 (1867): 569.

29. In summarizing Nicholas Dames's article on the Victorian practice of prolonged excepts, Rachel Ablow suggests the following: "Rather than offering the 'see, it works this way' epistemology of close reading, excerpt functions in Victorian reviewing and novel theory as 'see, it feels this way'"; Rachel Ablow, introduction to *The Feeling of Reading,* 5.

30. Dames, "On Not Close Reading," 13.

31. Oliphant, *Blackwood's Edinburgh Magazine,* 571. Also see Dames for a longer analysis of Oliphant's critical strategies. "What Oliphant intends to evoke," suggests Dames, "is the sheer feeling of Collins's novel." Dames, "On Not Close Reading," 18.

32. Oliphant, *Blackwood's Edinburgh Magazine,* 568, 572.

33. Edward Fitzgerald, "Letter to E.F. Pollock, December 1851," in *Wilkie Collins: The Critical Heritage,* ed. Norman Page (London: Routledge and Kegan Paul, 1974), 125.

34. Adams, *The Notting Hill Mystery,* 67–68. Page numbers for subsequent citations will be given parenthetically in the text.

35. Nicholas Dames, "Wave-Theories and Affective Physiologies: The Cognitive Strain in Victorian Novel Theories," *Victorian Studies* 46, no. 2 (2004): 210.

36. Paul Collins, "The Case of the First Mystery Novelist," *New York Times Sunday Book Review,* January 7, 2011, http://www.nytimes.com/2011/01/09/books/review/Collins-t.html?_r=0.

37. "Novelettes," *London Review* 11 (1865): 178.

38. "New Novels," *London Standard* 25 (May 1865): 3.

39. Ibid.

40. For instance, in reviewing *The Moonstone,* Geraldine Jewsbury writes, "Those readers who have followed the fortunes of the mysterious *Moonstone* for many weeks, as it has appeared in tantalizing portions, will of course throw themselves headlong upon the latter portion [of the novel] now that the end is really come, and devour it without rest or pause." Geraldine Jewsbury, *Athenaeum* 2126 (July 25, 1868): 106.

41. "New Novels," 3.

42. I refer to Dames's formulation of what he describes as "physiological novel theory," or a Victorian approach to novel reading that theorized it as "a performance—a performance enacted in and by the nerves." Dames, *Physiology of the Novel,* 11.

CHAPTER 8

Immoral Science in *The Picture of Dorian Gray*

Suzanne Raitt

Near the beginning of *The Picture of Dorian Gray*, the painter Basil Hallward explains to Lord Henry Wotton exactly what it is about Dorian Gray that inspired him to paint such an exquisitely beautiful portrait. Basil explains, "[Dorian] defines for me the lines of a fresh school" of art, and his "personality has suggested to me an entirely new manner in art, an entirely new mode of style. I see things differently, I think of them differently. I can now recreate life in a way that was hidden from me before."[1] In the course of the narrative, Hallward's phrase "recreate life" turns out not to be simply a metaphor. After Dorian's wish that the picture might "grow old" while he himself remains "always young" (25), the picture literally "recreates" life, renewing Dorian's fading body and absorbing into itself the processes of biological and moral decay that would otherwise engulf the living man. The picture substitutes for Dorian's mortal body so that the biology of aging is expressed not in the man but in the image. The immortality of art—its arrest of time and change—is transferred to the flesh that in normal circumstances would droop and wither as the body made its inexorable way toward death.

This interchangeability between man and image, between the dynamic processes described by science and the eternal stasis so prized in portrait art, is the focus of this chapter. Nineteenth-century advances in biological science—most significantly the development of cell theory in the 1840s, of which Oscar Wilde was aware[2]—had helped Victorian scientists understand the series of reparative mechanisms in the body that are designed to handle the toxicity and the waste products of the systems that sustain life: most importantly, cell metabolism and cell division.[3] Of course, the result of these processes was not immortality (although Freud fantasized that it might have been), but longevity.

Such advances informed a new interest in intervening in the rhythm of waste and repair, so that repair would outstrip or balance waste for as long as possible, for example by maintaining the body's health, and minimizing fatigue. As Tim Armstrong has noted, the same principles of thrift and efficiency—making use of every available material, even when it appeared inert or useless—that were used in the management of cities also applied to human biology,[4] and the "economy of the body," in Daniel Pick's words, became central to late Victorian social thought.[5] These biological models were enthusiastically taken up by writers on aesthetics as well as by scientists. Grant Allen, for example, novelist and aesthetician, used the concept in 1877 in *Physiological Aesthetics*, his inquiry into the nature of aesthetic pleasure: "The aesthetically beautiful is that which affords the Maximum of Stimulation with the Minimum of Fatigue or Waste."[6] Here art is aligned with the excess of stimulation over waste, or, as Allen put it earlier in *Physiological Aesthetics*, with "a state of high efficiency."[7] As Henry Adams put it, writing on the two laws of thermodynamics in 1910: "Matter indeed, is energy itself, and its economies first made organic life possible by thus correcting nature's tendency to waste."[8] Freud was fascinated by the idea of a world in which waste could be avoided and immortality guaranteed: "It may be . . . that [the] belief in the internal necessity of dying is only another of those illusions which we have created '*um die Schwere des Daseins zu ertragen*' [to bear the burden of existence]."[9]

Oscar Wilde also participated in this ongoing discussion about biological processes of renewal and their social implications. In his novel, however, he imagined a substitutive economy in which *art* might compensate for the limits of biology, entirely repairing the damages of life in all its ugliness. In effect, art "recreates" vitality, like a kind of magic medicine—or, as Wilde termed it, like an "immoral" science (17). By the end of the novel, this substitutive economy is exposed as a fantasy, and what might have seemed to be the stains of sin are revealed to be the inescapable marks of old age. Boldly, the conclusion posits biology as stronger than art; and art itself, when misapplied in this substitutive way, is exposed as merely a form of "immoral" science.

Wilde and Science

In recent years, a number of critics have explored the extent of Oscar Wilde's interest in science and suggested that scientific language, concepts, and discoveries played a significant role in the evolution of his art. Because so much of his life and work was devoted to defending aesthetic values, it might at first seem counterintuitive to think of Wilde in a scientific context. As Rita Felski has noted, however, "Though disdaining the rationalist claim of science, aestheticism was nevertheless deeply suffused by its organicist and pathological metaphors and by Darwinian notions of evolutionary development."[10] Further, as Philip E. Smith II has pointed out, "Wilde's knowledge of nineteenth-century science has been underestimated or misunderstood" by the majority of critics.[11] Michael Wainwright has claimed that "contradictory but contemporary scientific hypotheses informed Wilde's artistic practice" and has shown that *The Picture of Dorian Gray*, in particular, is built around competing theories of heredity.[12] Similarly, Carolyn Lesjak, examining the affinities between nineteenth-century atomic theory and *The Picture of Dorian Gray*, suggests that "nineteenth-century scientific thinking (and not just evolutionary biology), both in its methodological procedures and in its actual content, animates Wilde's aesthetic" and that "science as a discourse is very much at the center of the novel."[13] Implicit in all these critical analyses is the idea that Wilde's writing is deeply informed and shaped by his knowledge of science. I shall suggest, however, that *The Picture of Dorian Gray* goes even further than participating in a scientific discourse—it actually suggests that art is in fact a kind of science—but not one on which we should depend.

Wilde's early intellectual life was full of scientific inquiry. As Wainwright has shown, education at Oxford in the mid to late nineteenth century was newly focused on emerging sciences such as physiology and biology.[14] Wilde's commonplace book from his early years in Oxford in the 1870s reveals an interest in both psychology and biology, and a belief that the two were intimately connected. As John Wilson Foster has observed, "Wilde's Oxford notebooks [reveal] a surprising pleasure in science."[15] In an early entry Wilde observes, "There can be no knowledge of human nature without knowledge of the Laws of Mind, (Psychology) nor of the Laws of Mind without knowledge of the Laws of Life (Biology). / The science of society then rests on the science of life: sociology on Biology."[16] The commonplace books cite the work of contemporary Victorian scientists, including that of the physicist John Tyndall and biologist T. H. Huxley. At this time he also explored the writings of Victorian biologist Herbert Spencer, whose work Wilde addresses in his student essay "Historical Criticism," which celebrates the "scientific method" above all others.[17] In fact, many of the ideas he explores in this early essay anticipate the central concerns of *The Picture of Dorian Gray*. The idea of determinism, for example, emerges in his reflection that "the very first requisite for any scientific conception of history is the doctrine of uniform sequence: in other words . . . that the past is the key of the future."[18] Wilde also addresses the idea of decay: "All created things are fated to decay—a principle which, though expressed in the terms of a mere metaphysical abstraction, is yet perhaps in its essence scientific."[19] Clearly, Wilde was deeply interested in the implications of scientific writing and the philosophy of science from an early age, and, as will be shown, this attraction later emerges in the implicit analogy he draws between the scientist and the artist in *The Picture of Dorian Gray*.

The overlap between science and art is also reflected in Wilde's early intellectual environment, particularly through his relationship with Walter Pater, which fostered his sense that science and art might share a common language. Wilde met Pater, a fellow of Brasenose College, when he was still a student at Oxford in the fall of 1877.[20] Wilde was already fascinated by Pater's 1873 *Studies in the History of the Renaissance* (in *De Profundis* he called it "that book which has had such strange influence over my life"), and there are numerous echoes of it throughout *The Picture of Dorian Gray*.[21] As Billie Inman has shown, Pater was heavily influenced by the language and concepts of Victorian physiology: he predicted in 1889 that "for many years to come" the "enterprise" of the English language "may well lie in the naturalisation of the vocabulary of science. . . . The literary artist, therefore, will be well aware of physical science; science also attaining, in its turn, its true literary ideal."[22] In the conclusion to *The Renaissance* (1868), Pater describes combinations of "natural elements to which science gives their names" as lying behind "birth and gesture and death and the springing of violets from the grave": "Our life is but the concurrence," Pater writes, "of forces parting sooner or later on their ways."[23] The language and concepts of science lay behind even Pater's aestheticism, and they informed Wilde's own exploration of the supreme art of self-realization in "The Soul of Man under Socialism" (1891). Art for Wilde had a transcendent value, but he also theorized that science could lead the individual man to perfection: "Now and then, in the course of the century, a great man of science, like Darwin; a great poet, like Keats; a fine critical spirit, like M. Renan; a supreme artist, like Flaubert, has been able to isolate himself . . . and so to realise the perfection of what was in him, to his own incomparable gain, and to the incomparable and lasting gain of the whole world."[24] Artist and scientist were, in Wilde's view, both engaged in the art of self-exploration, self-actualization, and experimentation.

Science in Dorian Gray

The Picture of Dorian Gray introduces the reader to two scientist/artist figures: Dorian and Lord Henry. The latter, like the former, is neither an artist nor a scientist, but it is through him that the novel's discourse about influence as an artistic and scientific process is introduced and explicitly developed. In terms of his artistic prowess, Lord Henry thinks of himself as a kind of sculptor of the emotions:

> There was something terribly enthralling in the exercise of influence. To project one's soul into some gracious form, and let it tarry there for a moment . . . [Dorian] was a marvelous type, too . . . or could be fashioned into a marvelous type, at any rate. Grace was his, and the white purity of boyhood, and beauty such as old Greek marbles kept for us. There was nothing that one could not do with him. He could be made a Titan or a toy. (35–36)

Lord Henry imagines himself transforming Dorian as if he were a piece of marble, fusing with him in the process of re-creating him. But influence does not only flow one way. Musing on his relationship with Dorian, Lord Henry thinks: "There was something terribly enthralling in the exercise of influence. No other activity was like it. To project one's soul into some gracious form, and let it tarry there for a moment" (35); "To influence a person is to give him one's own soul" (17). His plan is "to dominate [Dorian]—[he] had already, indeed, half done so. He would make that wonderful spirit his own" (36). The ambiguity in the last phrase—making Dorian's spirit "his own" figures the young man both as a possession and as a quality (Dorian's spirit is also Lord Henry's spirit), and also signals the extent to which artistic creation, attraction, and mutual absorption are aligned in this novel. To be an artist is to create the thing you love, and in creating it, to change both it and oneself.

In addition to being an artist, however, Lord Henry occupies the role of what the novel terms an "immoral" scientist (17)—that is, a scientist whose experiments change not only the subject, but also the experimenter. Lord Henry thinks of his influence over Dorian as an art (he sculpts his spirit and in the process makes it his own), but he also describes it as a scientific experiment: "It was clear to him that the experimental method was the only method by which one could arrive at any scientific analysis of the passions; and certainly Dorian Gray was a subject made to his hand, and seemed to promise rich and fruitful results" (58). In this scientific context, however, his relationship with Dorian is figured as distinctly immoral. Lord Henry later explains that influencing and merging with another person, as he does Dorian, is ethically suspect: "All influence is immoral—immoral from the scientific point of view. . . . The aim of life is self-development" (17). Lord Henry's experiment on Dorian falls short of his scientific ideals, then, precisely because "the lad was his own creation" (57), not only an experiment but also the artistic product of a relationship. The artist fuses with his art in the process of transforming it, but the "moral" scientist scrupulously avoids such contaminating influence. In immoral science, transforming the nature of the object of the experiment cannot be distinguished from the transformation of the scientist himself. In the relationship between Lord Henry and Dorian, art is the instrument of immoral science, and Lord Henry is its inaugural practitioner.

Dorian, like Lord Henry, also conflates art with science. Enamored of his own image, he turns himself into a portrait by wishing that the painting, rather than his body, might undergo the indignities of aging. When he first gazes at his own picture, he fears not the effects of sin but the effects of time on his beauty: "There would be a day when his face would be wrinkled and wizen, his eyes dim and colourless, the grace of his figure broken and deformed. The scarlet would pass away from his lips and the gold steal from his hair" (25). In response, he becomes the unchanging image he so loves, while the portrait suffers the ravages of decay: "Hour by hour, and week by week, the thing upon the canvas was growing old. It might escape the hideousness of sin, but the hideousness of age was in store for it" (122). The painting, then, both expresses and contains the biological wasting processes of the body, while Dorian embarks on a life undertaken as art.

However, like Lord Henry, Dorian also sees himself as a scientist who must manage the effects of an experiment gone horribly awry. He views himself as having unwittingly unleashed a complex scientific process, toying with the notion that his relationship with the picture is owing to some strange quirk of biochemistry. He wonders at one point whether there may not be "some curious scientific reason" for what he calls the "horrible sympathy between him and the picture" (106). Wondering whether or not to pray that the "horrible sympathy . . . might cease," Dorian muses: "Was it really under his control? Had it indeed been prayer that had produced the substitution? Might there not be some curious scientific reason for it all? If thought could exercise its influence upon a living organism, might not thought exercise an influence upon dead and organic things?" (106). Dorian wonders if his effect on the picture was the result of a scientific experiment in which he, unwittingly, had exercised "influence" upon something inorganic. Lord Henry's self-styled "experiment" in influencing Dorian transforms flesh into art, body into marble, but Dorian's "scientific" undertaking works the other way: he exerts an influence on art that turns it into flesh, the painting into his own body. Lord Henry complains that to influence someone is to "give him one's own soul" (17), as he fears doing in the experiment and artistic creation that is his relationship with Dorian. If Lord Henry fears that the intensity of his influence over Dorian will somehow fuse the two of them, the danger for Dorian is even more menacing. In influencing the painting into its "horrible sympathy" (106) with himself, he literally "give[s] [his] own soul!" (17) As another immoral scientist, who, like Lord Henry, works primarily through influence, Dorian discovers soon enough that "when we thought we were experimenting on others we were really experimenting on ourselves" (59). When Dorian views the altered painting after the death of Sybil Vane, he feels that "his own soul was looking out at him from the canvas and calling him to judgement" (119). Dorian is simultaneously artist, scientist and experiment, instigator and victim of his desire for self-substitution.

The parallels Wilde draws in the novel between scientist and artist depend largely on the assumption that art and science share a conceptual framework. It is no coincidence, then, that the paradigm through which *Dorian Gray* explores the overlapping economies of "immoral" science and art originates in nineteenth-century understandings of the biology of human life. The rhythm of "waste and repair" that shapes Dorian's fantasy, with the picture repairing the waste of Dorian's aging body, was one of the key concepts in the new Victorian science of cell biology. Nineteenth-century scientists routinely refer to the rhythm of "waste and repair," or "waste and assimilation," which sustains life. Even as cells "waste," or wear out, new cells take over to repair the damage to the tissue, and the resulting equilibrium is essentially the biology of life. Herbert Spencer—one of the scientists whose work Wilde read at

Oxford—was one of the first to identify this rhythm, writing in 1864, "Repair is everywhere and always making up for waste."[25] Physician James Deane echoed him in 1869: "We have constantly in every human body a continual system of waste on the one hand, and on the other hand we have a perfect system of supply, going hand in hand together through all the stages of human life."[26] In 1900 E. B. Rosa wrote that the body "builds itself up and repairs waste."[27] Death was believed to result when the processes of assimilation or of repair could no longer keep up with the production of biological waste, a failure that was seen as inevitable in every living organism. As psychologist Henry Maudsley put it: "The common law of life is slow acquisition, equilibrium for a time, then a gentle decline that soon becomes a rapid decay, and finally death."[28] Pater also invokes the paradigm, writing in the conclusion to *The Renaissance,* a book that, as we have seen, Wilde loved, of the "perpetual motion" of the human body: "The passage of the blood, the waste and repairing of the lenses of the eye."[29] Here the body becomes an image for unceasing and self-sustaining movement, what Pater calls "that strange, perpetual, weaving and unweaving of ourselves."[30] Just like the world, the body constantly makes and unmakes itself such that it becomes an image for both ephemerality and longevity, both transient and resilient.

As Norton Wise and Crosbie Smith have explained, "the discourse of work and waste" was central to late Victorian culture and its anxieties about decay and degeneration.[31] Most of *Dorian Gray*'s early reviewers also participated in this discourse, curiously extending the reparative metaphor beyond the fictional narrative, even while they remained skeptical of the novel's fantasy of reparation. Over and over again critics used terms such as "filth," "muck," and "decay" in their condemnation of the novel. Samuel Henry Jeyes, for example, wrote in *St James's Gazette*: "Not being curious in ordure, and not wishing to offend the nostrils of decent persons, we do not propose to analyse *The Picture of Dorian Gray*," adding that the text "draws its life from malodorous putrefaction" and "delights in dirtiness."[32] An unsigned review in the *Daily Chronicle* called it "a poisonous book . . . heavy with the mephitic odours of moral and spiritual putrefaction," and another notice in the *Scots Observer* asks: "Why go grubbing in muck-heaps?"[33] Such language was informed by the pervasive anxiety about entropy and degeneration at the end of the century, which encouraged many social commentators to think of certain social groups—homosexuals among them—as themselves a form of waste or ordure. Cultural critic Max Nordau, for example, saw criminals, the insane, homosexuals, artists and city dwellers as the "refuse of civilized peoples"; French physician Charles Féré referred in 1888 to the "impotent, the mad, criminals or decadents of every form" as "the waste-matter of adaptation"; and journalist F. A. McKenzie referred to "waste humanity," as if certain people were somehow themselves a kind of garbage, a sign of the inexorable wasting away of the world.[34] Henry Adams noted that humanity is the most wasteful of all the species: "Man does more to dissipate and waste nature's economies than all the rest of animal and vegetable life has ever done to save them."[35] The stigmatization of waste as a sort of universal pathology appears in turn-of-the-century psychoanalysis, especially Freud's theory of sexuality, where a vision of a productive, teleological sexuality is haunted, as Leo Bersani has argued, by the promise of a masochistic self-shattering.[36] In vilifying the book in these terms, then, reviewers were implicitly expressing their own revulsion not just at Wilde's art, but also at his body and its habits.[37]

Immoral Immortality

Unlike the critics, who associated the novel with waste and found it repulsive as such, the narrative of *The Picture of Dorian Gray* communicates a fascination with waste and extravagance, all the while balancing the illicit nature of this morbid fascination with the apparent "moral" of the book. Dorian's self-destructiveness intensifies as he becomes increasingly (and misguidedly) desperate in his pursuit of beauty after the portrait starts to change, and he begins to realize that the equilibrium between waste and repair that he had hoped to set up is ultimately unsustainable. His first action, after he makes his prayer to the portrait, is on the face of it entirely harmless: he falls in love with Sybil Vane. Lord Henry sees this as the beginning of Dorian's transformation: "Lord Henry watched him with a subtle sense of pleasure. How different he was now from the shy, frightened boy he had met in Basil Hallward's studio! His nature had developed like a flower, had borne blossoms of scarlet flame" (54–55). Dorian's search for sensation becomes perverse (and the picture starts to change) only after he abandons Sybil. The urban landscape through which he walks when he leaves her signifies his descent into a more sinister world, in which destructive (and possibly homosexual) forms of pleasure are conflated with working-class and slum life:

> Where he went to he hardly knew. He remembered wandering through dimly-lit streets, past gaunt black-shadowed archways and

> evil-looking houses. Women with hoarse voices and harsh laughter had called after him. Drunkards had reeled by cursing, and chattering to themselves like monstrous apes. He had seen grotesque children huddled upon doorsteps, and heard shrieks and oaths from gloomy courts. (88)

Here a classic 1890s depiction of nightmarish streets and degraded people is used to imply other, darker forms of pleasure than those that Dorian has hitherto explored. Although here and elsewhere Wilde's decadent prose aestheticizes the ugliness he embraces, the narrative also emphasizes that Dorian's gradual immersion in what he at first identifies as a new type of pleasure is in fact a regressive move into more primitive, even bestial modes of being. The "Hellenic ideal," advocated by Lord Henry at the opening of the novel (18), eventually becomes the "New Hedonism" (22), and Dorian's search for beauty becomes indistinguishable from a willed self-corruption.

The picture itself not only tracks this change but also starts to stimulate Dorian to seek his own decay. When he compares the wizened portrait to the beauty he sees in the mirror, the "very sharpness of the contrast used to quicken his sense of pleasure. He grew more and more enamoured of his own beauty, more and more interested in the corruption of his own soul" (128). The picture itself becomes a kind of addiction. In his fascination with the influence he can exert on this image of himself, Dorian starts deliberately to seek out sensations that will lead to self-transformation: "In his search for sensations that would be at once new and delightful, and possess that element of strangeness that is so essential to romance, he would often adopt certain modes of thought that he knew to be really alien to his nature, abandon himself to their subtle influences" (132). In changing himself, he also changes the portrait, experimenting endlessly and then returning to view the results of his sins: "He would sit in front of the picture, sometimes loathing it and himself, but filled, at other times, with that pride of individualism that is half the fascination of sin, and smiling with secret pleasure at the misshapen shadow that had to bear the burden that should have been his own" (140). Dorian's obsession with his influence over the picture to which, in some "curious scientific" manner (106), he has given his soul, exposes the immorality of his scientific endeavors, since influence, as Lord Henry explains, is incompatible with science (17).

If, as I have suggested, the economy of Dorian's experiment on himself is modeled on the biology of life, then Dorian's tragic end reflects the gradual, but inevitable, decline of the body's ability to protect itself against its own waste. In Tim Armstrong's words, "'Waste,' like fatigue, signals the point at which the body and the machine cannot readily be reconciled."[38] Experiments on unicellular organisms in the 1890s had already confirmed the insight of doctors such as Henry Maudsley: "The products of organic decomposition are fatal to the organism, if not eliminated or counteracted, and the most virulent and fatal [are] those that are derived from the corruption of its own substance."[39] Similarly, the portrait, which was designed to absorb the decay of Dorian's depraved flesh, begins to remind him of what he had sought to repress. It becomes the embodiment of the waste his body has expelled but not destroyed, and thus becomes dangerous to the organism that paradoxically relies on its processes. The painting's "changing features showed him the real degradation of his life," and Dorian's successive fascinations with perfumes, music, jewels, embroidery, and ecclesiastical vestments are, as the novel tells us, merely "modes by which he could escape, for a season, from the fear that seemed to him at times almost too great to be borne" (140). When the picture ceases to function as reparation and starts to become a reminder of guilt and mortality, Dorian quickly descends into paranoia, cycling rapidly through his series of obsessions as if he were trying to outrun his own inevitable decline: "He hated to be separated from the picture that was such a part of his life, and was also afraid that during his absence some one might gain access to the room, in spite of the elaborate bars that he had caused to be placed upon the door" (141). Far from being a protective mechanism, the portrait ultimately becomes the material trace of his inner and outer degradation, an image of what must be repudiated and expelled in order to sustain life. But Dorian, as artist and "immoral" scientist, cannot repudiate it, precisely because he is fused with the image he created, coextensive with his own experiment.

Tellingly, at this critical juncture a conventional scientist, in the shape of Alan Campbell, comes to the rescue. Campbell apparently performs a sort of miracle through science—to make Dorian's most glaring by-product of moral waste—Basil Hallward's murdered body—literally disappear, presumably through the application of vaporizing chemicals. Dorian, desperate to destroy Hallward's corpse, hails Campbell with relief: "Alan, you are scientific. You know about chemistry and things of that kind. You have made experiments. . . . All I ask of you is to perform a certain scientific experiment" (168). Dorian expects that Campbell will do the experiment without "turn[ing] a hair" (169), since for Campbell, the practice of science is not about the scientist himself (as it is for Dorian), but about "increasing the sum of knowledge in the world, or gratifying intellectual curiosity, or something of that kind" (170)—or at least so Dorian would like to think. And although Campbell is "pale" when he finally returns from performing his ghastly work, he is "absolutely calm" (174). The horror he expresses in his subsequent suicide seems to derive as much from his discovery that Dorian could expose something dreadful in his past, and from his reluctance to have anything to do with Dorian, as from his experiences in the attic room with Basil Hallward's body (170–71). The strange science of Dorian's relationship to the portrait may have its uses, but conventional science—and the biology of mortality with which, in this novel, it is associated—win out in the end.

In *Dorian Gray*, then, we see the fantasy that art, when used as an "immoral" science, might prolong life. But that fantasy cannot last forever. The nineteenth-century scientists of waste and repair cited by Freud in his investigation of whether death is inevitable for all living things found that single-celled organisms could survive indefinitely only if they were protected from their own waste. Lorande Woodruff, professor of biology at Yale in the late 1800s, found that the "'slipper-animalcule,' which reproduces by fission into two individuals, persisted until at least the 3029th generation."[40] According to Freud, Woodruff was only able to obtain these startling results by continually providing fresh nutrients to each generation. Freud concluded that "if it is left to itself, [the animalcule] dies a natural death owing to its incomplete voidance of the products of its own metabolism."[41]

Similarly, in the novel, the portrait's mechanisms eventually prove inadequate to the task. Instead of being invigorated by looking at the portrait, by the end of the novel Dorian feels only fear when he thinks of it. Instead of protecting him, it seems to threaten him, to gather up all the detritus of his history and to mock him with it. Its very existence makes him vulnerable to exposure: "There was only one bit of evidence left against him. The picture itself—that was evidence. He would destroy it. Why had he kept it so long? Once it had given him pleasure to watch it changing and growing old. Of late he had felt no such pleasure. It had kept him awake at night. When he had been away, he had been filled with terror lest other eyes should look upon it" (222–23). Dorian is finally destroyed by what he can neither assimilate nor escape, the very by-products of his hateful and wasteful life.

Recontextualizing the novel in the discourse of nineteenth-century science suggests that it is about the inexorability of scientific truths as much as about the self-destructive nature of pleasure. After all, the changes to the portrait reflect not just Dorian's malevolence, but also the transformations of age: "It had altered already, and would alter more. Its gold would wither into grey. Its red and white roses would die" (91–92). As Ellie Ragland-Sullivan notes, the portrait is not just "an allegorical depiction of an ethical state," but also "a caricatured picture of old age, seen from the slant of a skewed narcissism."[42] This picture of old age is not just disturbing for psychological reasons, it is also horrifying for existential reasons, serving as a metaphor for the inexorable nature of biological decay. Contemplating the picture, Dorian wonders "which were the most horrible, the signs of sin or the signs of age" (128). Significantly, Dorian is not just a bad man, he is a bad *old* man, and his final attack on the painting is an expression of horror at his bodily decline as much as his moral failings: "It was his beauty that had ruined him, his beauty and the youth that he had prayed for" (220). Hidden in what is apparently a profoundly—if perversely—moral tale is an impotent rage against the inexorability of scientific reasoning and its results, and the impotence of art to protect against them. Waste and repair might be the rhythm of life; but eventually they become the signposts of death.

Notes

1. Oscar Wilde, *The Picture of Dorian Gray* (1891; Oxford: Oxford University Press, 1981), 10. Subsequent references in the text are to this edition.

2. See Michael Wainwright, "Oscar Wilde, the Science of Heredity, and *The Picture of Dorian Gray*," *English Literature in Transition, 1880–1920* 54, no. 4 (2011): 495.

3. William Turner, writing in 1889 about the development of cell biology in the nineteenth century: "In 1839 Theodore Schwann published his famous researches into the structure of animals and plants, in which he announced the important generalization that the tissues of the animal body are composed of cells, or of materials derived from cells." Turner, "The Cell Theory, Past and Present," *Journal of Anatomy and Physiology* 24 (1890): 257.

4. Tim Armstrong, *Modernism, Technology, and the Body: A Cultural Study* (Cambridge: Cambridge University Press, 1998), 43.

5. Daniel Pick, *Faces of Degeneration: A European Disorder, c.1848–c.1918* (Cambridge: Cambridge University Press, 1989), 6.

6. Grant Allen, *Physiological Aesthetics* (London: Henry S. King, 1877), 39.

7. Ibid., 32.

8. Henry Adams, "A Letter to American Teachers of History," in *The Degradation of the Democratic Dogma* (1910; repr. New York: Macmillan, 1919), 215.

9. Sigmund Freud, *Beyond the Pleasure Principle*, trans. James Strachey (1920; New York: Norton, 1989), 53.

10. Rita Felski, "The Counterdiscourse of the Feminine in Three Texts by Wilde, Huysmans, and Sacher-Masoch," *PMLA* 106, no. 5 (1991): 1098.

11. Philip E. Smith II, "Protoplasmic Hierarchy and Philosophical Harmony: Science and Hegelian Aesthetics in Oscar Wilde's Notebooks," in *Critical Essays on Oscar Wilde*, ed. Regenia Gagnier (New York: Prentice Hall, 1993), 203.

12. Wainwright, "Oscar Wilde," 494.

13. Carolyn Lesjak, "Oscar Wilde and the Art/Work of Atoms," *Studies in the Literary Imagination*, 43, no. 1 (2010): 5, 15.

14. Wainwright, "Oscar Wilde," 494–95.

15. John Wilson Foster, "Against Nature? Science and Oscar Wilde," *University of Toronto Quarterly* 63, no. 2 (1993–94): 332.

16. Wilde, *Oscar Wilde's Oxford Notebooks: A Portrait of Mind in the Making*, ed. Philip E. Smith II and Michael S. Helfand (Oxford: Oxford University Press, 1989), 109.

17. Wilde, "Historical Criticism," in *The Complete Works of Oscar Wilde*, vol. 4, ed. Josephine Guy (Oxford: Oxford University Press, 2007), 47.

18. Ibid., 28.

19. Ibid., 31.

20. Richard Ellmann, *Oscar Wilde* (Harmondsworth: Penguin, 1988), 80.

21. Wilde, *De Profundis*, in *De Profundis and Other Writings*, ed. Hesketh Pearson (London: Penguin, 1987), 158.

22. Billie Andrew Inman, *Walter Pater's Reading: A Bibliography of His Library Borrowings and Literary References, 1858–1873* (New York: Garland, 1981), 182ff.; Walter Pater, "Style," in *Essays on Literature and Art*, ed. Jennifer Uglow (London: Dent, 1973), 66.

23. Walter Pater, conclusion to *The Renaissance*, in *Three Major Texts*, ed. William E. Buckler (New York: New York University Press, 1986), 217, 218.

24. Wilde, "The Soul of Man" (1891), in *The Soul of Man and Prison Writings*, ed. Isobel Armstrong (Oxford: Oxford University Press, 1990), 1.

25. Herbert Spencer, *The Principles of Biology*, 2 vols. (London: Williams and Norgate, 1864), 1:171.

26. James Deane, *An Essay on the Waste and Supply in the Human System* (London: Longmans, Green, Reader, and Dyer, 1869), 56.

27. E. B. Rosa, "The Human Body as an Engine," *Popular Science Monthly* 57 (1900): 496.

28. Henry Maudsley, *Body and Will* (London: Kegan Paul, Trench, 1883), 319.

29. Pater, conclusion to *The Renaissance*, 217.

30. Ibid., 219.

31. M. N. Wise, with the collaboration of Crosbie Smith, "Work and Waste: Political Economy and Natural Philosophy in Nineteenth Century Britain," pt. 2, *History of Science* 27 (1989): 421.

32. Samuel Henry Jeyes, "A Study in Puppydom," in *Oscar Wilde: The Critical Heritage*, ed. Karl Beckson (London: Routledge and Kegan Paul, 1970), 68, 72, first published in *St James's Gazette* 20 (1890): 3–4.

33. Unsigned reviews of *The Picture of Dorian Gray* in Beckson, *Oscar Wilde*, first published in the *Daily Chronicle*, June 30, 1890; and *Scots Observer*, July 5, 1890: 72, 75.

34. Max Nordau, *Degeneration* (London: Heinemann, 1895), 337; Charles Féré, quoted in Pick, *Faces of Degeneration*, 32; F. A. McKenzie, *Waste Humanity: Being a Review of Part of the Social Operations of the Salvation Army in Great Britain* (London: Salvation Army, 1908–9), 20.

35. Adams, "A Letter," 216.

36. Leo Bersani, *The Freudian Body: Psychoanalysis and Art* (New York: Columbia University Press, 1986), 29–50.

37. Many of the people who were identified as human refuse, however, fought back by reveling in their own liberation from the rigors of work and thrift. Wilde writes about this class in "The Soul of Man," describing them as "real men": "They are either under no necessity to work for their living, or are enabled to choose the sphere of activity that is really congenial to them, and gives them pleasure. These are the poets, the philosophers, the men of science, the men of culture—in a word, the real men" (3).

38. Armstrong, *Modernism, Technology*, 65.

39. Maudsley, *Body and Will*, 322.

40. See Freud, *Beyond the Pleasure Principle*, 56–57.

41. Ibid., 58.

42. Ellie Ragland-Sullivan, "The Phenomenon of Aging in Oscar Wilde's *The Picture of Dorian Gray*: A Lacanian View," in *Memory and Desire: Aging in Literature and Psychoanalysis*, ed. Kathleen Woodward and Murray M. Schwartz (Bloomington: Indiana University Press, 1986), 119, 124.

PART III

Strange Energies

Reconceptualizing the Physical Universe

CHAPTER 9

Chaotic Fictions

Nonlinear Effects in Victorian Science and Literature

Barri J. Gold

In his 1853 preface to *Bleak House,* Charles Dickens responds to two objections to the novel: the first objection comes from a Chancery judge who claims that the operations of Chancery (the central object of Dickens's sustained satire) are "immaculate," its progress impeded only by a "parsimonious public" unwilling to allow the appointment of more judges.[1] The second objection comes from more or less scientific authorities, who insist that spontaneous combustion, such as that which kills the novel's Mr. Krook (chap. 32), is simply impossible. Dickens's response to these objections constitutes the whole of his preface, in which he insists that everything set forth in the pages that follow regarding Chancery is "substantially true" (preface, 5). He insists, moreover, that his portrayal of spontaneous combustion rests on the testimony of numerous reliable authorities of his own. Only his final caveat suggests that he may have distorted reality for the sake of good fiction: "In *Bleak House* I have purposely dwelt upon the romantic side of familiar things" (preface, 7).

Dickens implies that "familiar things" actually have a "romantic side," that the portrayal of such a romantic side may be "substantially true," and, moreover, that this kind of truth inheres both in the physical world and in the social. What he means by "romantic," here, is not entirely clear. It seems rather unlikely that *Bleak House* is intended to portray anything along the lines of an idealized reality. It makes more sense to consider Dickens as alluding to Romantic emphases on emotion and intuition in the face of an Enlightenment rationalization, especially of nature. But Dickens quite clearly does not set the romantic in opposition to the realistic; the romantic, that which we feel or imagine but cannot confirm, far from being opposed to the empirical, even positivist, mandate we associate with realism, is part and parcel of the real.

Needless to say, the spontaneous combustion of Mr. Krook has been much discussed over the years. Contemporary scholars have interpreted the incident in numerous ways: as solely metaphorical, as testing our willingness to suspend disbelief, and as representing Dickens's resistance to scientific authority.[2] As one critic put it, "In the end, most of us agree that, for literary purposes, the scientific accuracy of Spontaneous Combustion doesn't matter."[3] And yet we can't leave it alone. In this we join a long and proud tradition: Krook's smoldering embers immediately sparked an extended conversation between Dickens and his friend George Henry Lewes, who bemoaned the inclusion of the scene because he thought it likely to set back public scientific knowledge considerably, "tend[ing] to perpetuate the error in spite of the labours of a thousand philosophers."[4] While Dickens fails to establish the truth of spontaneous combustion, his efforts to do so make visible the process by which fiction may participate in transforming statements into facts. The failure to thus factify spontaneous combustion, on the one hand, illustrates that such transformations cannot be effected by fiction or even language alone. On the other hand, this failure brings into focus the contributions *Bleak House* makes toward establishing entropy as a thing and the second law of thermodynamics as a fact. While discursive elements are essential to the transformation of statement into fact and to the shaping of the scientific object, the material world must cooperate. Thus, "the immense popularity of [our] dear Dickens" that so worries Lewes lest the ignorant be persuaded, can do much to promote the acceptance and further dissemination of the principles of entropy that *Bleak House* depicts so compellingly but, for reasons of chronology, cannot name.[5] In this case, discourse cooperates with the universe to establish entropy as a scientific object.[6]

Although spontaneous combustion often functions as an intellectual scapegoat that carries away the sins of scientific inaccuracy, Dickens may not be so far off in his thinking as he first appears, either from the science of his day or from that of our own.[7] His nonlinear reasoning is indicative of certain tendencies in Victorian scientific fictions more broadly: those flights of nonlinear fancy in both fictional and scientific writing that wrestle with the myriad natural systems whose apparent randomness, disproportionate effects, and unpredictability baffle traditional mathematics and science. In what follows, I gesture to a few of what I call *chaotic* fictions in Victorian literature and science. Such fictions (a word I construe quite broadly) may be found in the writings of authors as different as Alfred, Lord Tennyson, Norman Lockyer (astronomer and friend of Tennyson), Herbert Spencer, James Prescott Joule, William Thomson (later Lord Kelvin), and his less well-known colleagues Balfour Stewart and P. G. Tait. These fictions, I wish to argue, represent more than a mere mash-up of Victorian science and fancy. Instead, they are indicative of a Victorian attitude toward scientific speculation, especially within the science of energy and its "North British School," with which the above-named physicists were associated.[8] Here we find a peculiar mix of the romantic, the religious, and the scientific, the connections between which, Stewart and Tait insist, have foundation in the quintessentially scientific principle of continuity. Where continuity is assumed, principles flow across length and time, even across disciplinary boundaries, such that the novelist and the scientist converge in their social and moral, as well as scientific, speculations. And continuity assures us that even the most perplexing phenomena must have a familiar side—or, as George Levine observes of Dickens, "The ordinary . . . is latent with possibilities of the extraordinary."[9]

Hidden Order

Driven by a deeply held hope that order and purpose were at the base of even the most dismaying of natural phenomena, the Victorians sought for "the hidden order that exists within chaotic systems."[10] Such a hope suffuses Tennyson's *In Memoriam*; so popular was this elegy, so epigrammatic did its phrases become, that we may well take it as a litmus of Victorian concerns regarding the apparently chaotic, unpurposeful, and violent in nature. As I have argued elsewhere, such hopes—evident in Tennyson's "faintly trust[ing] the larger hope" that in spite of appearances, "God and Nature" might not be "at strife"—were indeed critical in shaping the Victorian science of energy.[11] The impulse to reason away destruction and waste similarly emerges in James Prescott Joule's earliest statements about the conservation of energy. Despite evidence to the contrary, Joule is firm in his belief that it would be "manifestly absurd to suppose that the powers with which God has endowed matter can be destroyed any more than they can be created by man's agency."[12]

Such a belief is also foundational in the counterentropic transformation Tennyson articulates in imagining the birth of a child at the end of *In Memoriam.* Moving from the language of waste that dominates the earlier parts of the poem to the etymologically linked concept of vastness, Tennyson imagines that "star and system rolling past, / A soul shall draw from out the vast /And strike his being into bounds" (CXXXI, 291). Such an image resonates with Dorian Gray's late-century fantasy of awakening to "a world that had been re-fashioned anew in the darkness for our pleasure."[13] And even Herbert Spencer, driven by his wish to establish the possibility of renewal in nature in spite of the drive to universal entropy, posits a mechanism for the restoration of order from a universe, or at least a solar system, that has gone to its final rest: "Certain of the great facts which science has established imply potential renewals of life, now in one region now in another, followed, possibly, at a period unimaginably remote by a more general renewal."[14] Chaos thus figures "as order's precursor and partner, rather than as its opposite,"[15] something understood (as Tom Stoppard observes regarding the second law of thermodynamics) "by poets and lunatics from time immemorial."[16] In such cases, it seems that only the science lags.

Tennyson's image of counterentropic transformation, moreover, soon gives way to a picture of human development that not only finds potential order in chaos, but also suggests the likeness of processes across vastly different scales. In Tennyson's account, the unborn child "move[s] through life of lower phase" before it can be born and think and result in man (CXXXI, 291). Tennyson, of course, anticipates what would be termed "recapitulation theory," popularly disseminated in the phrase (generally attributed to Ernst Haeckel) "Ontogeny recapitulates phylogeny." His suggestion that the life of his friend plays out in miniature the development of the species, that the movement of the species "from more to more" (prologue, 203) echoes the development of the earth, suggests the recursion of complex patterns at vastly different scales of time and size. Tennyson, of course, did not have access to the term "fractal" (coined by the mathematician Benoit Mandlebrot in 1975). He uses, however, the term "type," not only as a noun, but also as a verb. As a noun, "type" in its Victorian conception gestures backward to a notion of the archetype, grounded in a view of an essentially stable natural world. The verb form, however, implies at once that change inheres in the soon-to-be Darwinian concept of species. And for all that recapitulation theory has been disproved, the notion that each individual undergoes in miniature the whole of evolution, that the development of man, in Tennyson, "type[s] this work of time" (CXVIII, 281), suggests a larger Victorian fascination with similarities across scale.

A particularly telling example occurs in a short story entitled "The Tree of Knowledge" published in 1853 in *Dublin University Magazine.* After noticing that the marks left by his electrical experiments bear "a strange and striking resemblance to the foliage of a tree, imitating with a marvelous mimicry" stems, branches, leaves, and sap down to the "most delicate extremities of the plant," the narrator proceeds to find the pattern everywhere.[17] Such repetition of patterns across systems and scales accounts for much that, according to contemporary standards, is "wrong" and a surprising amount that turns out to be "right" in Victorian science. Correct or incorrect, much of Victorian science could not have been carried on without the presumptions attached to the capacity of systems to type other systems. The geological principle of uniformitarianism that informs Tennyson's "type" was essential to Darwin's reasoning, as was the notion that what could be done on a small scale (artificial selection) was similar to what happened on the large (natural selection). On the other hand, William Thomson (later Lord Kelvin) famously calculated a far-too-short lifetime for the sun based on the presumption that it burned in the same way as "matter in our laboratory" only on a much larger scale—an assumption of similarity eventually undermined by the advent of nuclear physics.[18]

The physicist Balfour Stewart and the astronomer Norman Lockyer (also friends with Tennyson) take the notion of the type quite literally in their scientific speculations. Their two-part article "The Sun as a Type of the Material Universe" anticipates much of what is elaborated in Stewart's *The Conservation of Energy* as well as in *The Unseen Universe.*[19] The article begins with a discussion of solar physics, concluding that the molecular state of the sun must be one of "infinite delicacy" (327). They also observe that the "manifest relation" (327) between sunspots and the positions of both Venus and Jupiter suggests a relationship between the "different members of our system" and of the universe more broadly, far more intimate than even our mathematical calculations would suggest: "They feel, they throb together, they are pervaded by a principle of delicacy even as we are ourselves" (327). The article's title further underscores the authors' desire to find a likeness between systems of vastly different scale. For the sun to figure as a "type" of the material universe is a kind of physically manifested synecdoche; its operations enact in (relative) miniature those of the larger system of which it is part. At the other end of the scale, solar activity figures as a very large model of our own smaller systems, enacting macroscopically the submicroscopic principles at the roots of life itself, for it is among the goals of their essay to argue for "the place of life in a universe of energy" (319).

Stewart and Lockyer further identify this similarity across scales as necessarily connecting disciplines:

> There is often a striking likeness between principles which nevertheless belong to very different departments of knowledge. Each branch of the tree of knowledge bears its own precious fruit and there is a unity in this variety—a community of type throughout. Nor is this resemblance a merely fanciful one, or one which the mind conjures up for its own amusement. While it has produced a very plentiful crop of analogies, allegories, parables and proverbs, not always of the best kind, yet parables and proverbs are or ought to be not fictions but truths. (319)

They claim such "unity in variety" exists across such "different departments of knowledge" as biology, astronomy, energy physics, social relations, and imperial politics. Not mere analogy or even synecdoche, the part exhibits the properties of the whole in ways at least some Victorian scientists understood as "not fictions but truths."

Such "true" analogies shape Stewart and Lockyer's prose as they introduce their readers to energy both kinetic (the energy of motion) and potential (the energy of position) through an extended analogy to social energy. Like so many evocations of what's natural, their use of this trope reflects their political concerns. And like other applications of thermodynamic principles to social dynamics, theirs ring remarkably conservative, in a social as well as an energetic sense.[20] Nonetheless, the real analogies described by Stewart and Lockyer do not allow us to distinguish as clearly as we might like between literal and figurative likeness, because there is little distinction between likenesses of degree and likenesses of kind. When the authors suggest that "energy in the social world is well understood," or that breaking a chemical bond is like lifting a stone from the earth, or that "food is the fuel which we burn in our own bodies instead of on our hearths or in our engines," such real analogies suggest not a dichotomy, but rather a continuum between the literal and the figurative (321–23). Metaphorically related, these various processes are also different manifestations of the same scientific principles: the social, the gravitational, the chemical are all instances of the transformation of energy, governed by the laws of thermodynamics.

As they move from the social to the scientific, Stewart and Lockyer find likeness between physical systems at vastly different scales, between different departments of scientific knowledge, between chemistry and physics. It is thus no accident that in enumerating such histories of the "creation" of potential energy, they use the same language to discuss both gravitation and chemical bonding. They claim we obtain such usable energy "when we tear asunder a stone from the earth" or when we "tear asunder the component atoms of some chemical compound" (321). And while we might more exactly refer to the chemical potential or higher energy state that results from separating the atoms of carbonic acid, Stewart and Lockyer refer to this as "a very convenient form of energy of position," emphasizing the likeness between astronomic, atomic, and experiential-sized systems (321). They further emphasize that the effectiveness of such an energy-storing enterprise depends on "whether our scale of operations be sufficiently great" (321). Their "community of type" thus suggests a kind of self-similarity or scale invariance: physical systems repeat themselves, their shapes and processes, at very different scales, some as familiar as our day-to-day interactions, some as strange as the microscopic workings of the mind, the splitting of a molecule, or the production of sunspots.

Continuity

The notion that the familiar and the unfamiliar share such a "community of type" is consistent with the scientific principle of continuity. This principle—that what we learn or hypothesize about the natural world must be consistent with what we already know—constrains and shapes any paradigm shift within the sciences. For example, the quantum mechanical models of the early twentieth century, when taken to macroscopic limits, are constrained to agree with Newtonian predictions that are backed by over two centuries of observation and common sense. With this in mind, Joule's insistence that it would be absurd to suppose that man's agency could create or destroy what we come to call energy may be understood as his insistence that what he finds must be consistent with what he already knows. It then follows that when Joule and his contemporaries found so much evidence to the contrary (in the heat loss associated with all mechanical processes), they knew they hadn't yet got the whole picture; along with all that observable heat loss, there must be conservation as well.

In their book *The Unseen Universe,* Balfour Stewart and P. G. Tait put such commonsense reasoning to a rather different use. They wished to refute those who would find science opposed to religion, to "[strip] off the hideous mask with which materialism has covered the face of nature" (xv).[21] They tried to find the confidence that George Levine ascribes to Dickens in his portrayal of Krook's death, "the confidence of natural theology, in which material reality corresponds meaningfully to a moral reality."[22] Ultimately, they hoped to demonstrate that notions of an afterlife are consistent with the second law of thermodynamics. Tamara Ketabgian, whose chapter in this volume discusses this project far more fully, notes that in spite of the authors' insistence that their methods are "absolutely driven by scientific principles," *The Unseen Universe* provoked widespread scientific criticism and spiritualist emulation (see Ketabgian n. 7). What interests me here is a particular scientific principle by which they claim to be driven—the principle of continuity—and the various ways it enables their chaotic fictions.

The principle of continuity, generally attributed to Gottfried Leibniz (who, in addition to being a philosopher was also a mathematician best known for developing calculus independently of Newton), has been summed up as the notion that "nature never makes leaps."[23] Leibniz's own statement that "the rules of the finite are found to succeed in the infinite," reveals its close association with infinitesimal calculus, wherein we can add infinitesimally small pieces to determine the area delineated by complex curves.[24] In *The Unseen Universe,* Stewart and Tait, undoubtedly of a mind with Leibniz in his conviction that the principle should serve "not only as a test" of scientific theories (such as that applied to quantum mechanics, mentioned above), "but also as a very fruitful principle of discovery,"[25] complain that neither the "extreme scientific school," nor "the old theological school" have "loyally followed" this principle, which, they insist in words that echo Leibniz, "underlies not only all scientific inquiry, but all action of any kind in this world" (87, xx). They write:

> All this follows from the principle of Continuity, in virtue of which we make scientific progress in the knowledge of things, and which leads us, whatever state of things we contemplate, to look for its antecedent in some previous state of things also in the Universe. This principle

> represents the path from the known to the unknown, or to speak more precisely, our conviction that there is a path. (xv–xvi)

From this perspective, scientific progress is possible only when we can reconcile emerging scientific principles with knowledge of how the world is, or even our strongly held convictions of how it should be. The latter is tricky ground, of course, though it is key to Stewart and Tait's argument. In the state of our knowledge as well as in states of the physical world, before and after must connect in the present.

Victorians certainly were not proof against finding anticipations of their own science in the beliefs of the past. Stewart and Tait were quite explicit about the connections between ancient religion and contemporary science. They evoke the ancient Egyptians, whose records attest not only to belief in the immortality of the soul, but to the grounding of this belief in something very like the first law of thermodynamics—the law of conservation of energy: "Dissolution, according to them, is only the cause of reproduction—nothing perishes which has once existed, and things which appear to be destroyed only change their natures and pass into another form" (5). Such assertions are similarly prominent in the beliefs of Edward Bulwer-Lytton's fictional underground race, the Vril-ya, whose buildings evoke "the earliest form of Egyptian architecture . . . the Corinthian capital. . . . Etruscan vases [and] the walls of Eastern sepulchers" (24). These architectural impressions are reinforced by the figures, attire and demeanor—the Oriental quietude and sphinxlike faces—of the people, whose belief in an afterlife is grounded in a conviction not unlike Joule's: "They hold that wherever He has once given life, with the perceptions of that life, however faint it be, as in a plant, the life is never destroyed; it passes into new and improved forms, though not in this planet" (83).

However fanciful such connections may seem, we still find such intuition incredibly useful to scientific hypothesizing. This process resonates strongly with what Charles Sanders Peirce calls "abductive reasoning." Peirce's abduction is the step between perception and reasoning, a logical process that precedes deduction and induction. Far more recently, Wendy Wheeler has used Peircean semiotics as a model for the synthesis of science and the humanities that drives contemporary ecocriticism, with its willingness to embrace the poetic, the numinous, as well as scientific accounts of how the world works. Peircean semiotics sidesteps the proliferation of arbitrariness in meaning evolved from the Saussurean model, by rooting signs in the material world—signs we read long before our personal or phylological development of language. In this way, "New theories and models are [indeed, must be] forged from creative re-readings of the past."[26] Peirce, in other words, puts forth a model for what Stewart and Tait assure us there must be: a "path from the known to the unknown" (xv–xvi).

Delicate Constructions and Explosive Effects

The path from the known to the unknown articulated by the law of continuity and enabled by abductive reasoning is highly suggestive regarding the connections between the literal and figurative found above in the connections Stewart and Lockyer make among the actions of gravitation, of chemical attraction, and of "the force which keeps a man down in the world."[27] The principle of continuity suggests how these metaphorical connections might be part and parcel of the connectedness of the world. If Stewart and Lockyer oversimplify (and politicize) by ascribing one man's high position to the personal energy expended by the founder of the family, they are nevertheless more sensitive to the difficulties of tracing such causes in material phenomena. For this, they evoke Leibniz's continuity principle. Within their "principle of delicacy," however, they move not only from the known to the unknown, but from the finite to the infinite, as they identify systems of "great delicacy" and hypothesize systems of "infinite delicacy." Going a step further than their "community of type," their principle of delicacy implies that the typologically similar systems in question are similarly complex at every scale.

Great delicacy, for Stewart and Lockyer, accounts also for a common feature of such systems—that within them, small or even imperceptible causes may produce disproportionately large effects. They conclude that the sun is such a system, a star "of great delicacy, so that in our luminary a very small cause might be the parent of enormous effects, of a visible and mechanical nature" (327). Contemporary science would say that such a system exhibits sensitive dependence on initial conditions. We find it not only in astronomical phenomena, but also in the apparent unpredictability of global markets, as well as in daily disappointments with our local weatherman. Stewart and Lockyer, too, posit that systems of great delicacy exhibiting such sensitive dependence may be found, or at any rate imagined, at more familiar length-scales: "By an amount of directive energy less than any assignable quantity a current may be made to start suddenly, cross the Atlantic, and . . . explode a magazine on the other side" (325). A tiny spark in a gun with a hair trigger, poised just so, may win an empire.[28] A small change in the initial conditions—a shift in the position of the gun, an eddy in the winds of change—and perhaps nothing of note will occur. But poised just so, and the effects may well be, literally and figuratively, explosive.

As Stewart and Lockyer move from this relatively familiar machine of great delicacy to the less familiar workings of the mind, life, and even a Supreme Being, they posit machines of "infinite delicacy" whose unpredictability is nonetheless fully consistent with a physical universe governed by the principles of energy. Denying the possibility of a living being whose actions are fully calculable, they locate "the very perfection of . . . animated beings . . . in the fact that their motions cannot possibly be made the subject of calculation."[29] Such freedom, through which living beings not only produce motion but produce it discontinuously, must nonetheless conform to the laws that govern the physical universe. For this reason they propose that the delicacy of living beings enables them to supply an infinitely small amount of "directive energy" that nonetheless brings about perceptible results (326). Thus, even without the nonlinear mathematics that explains such disproportionate effects, they provide a model of how something may be rooted in physical causes but defy prediction by articulating a principle of sensitive dependence on initial conditions.

Stewart and Lockyer thus beat back the materialist monster, both "the man who could predict his own motion" and, presumably, the godless scientist who could conceive of such a horror (324). Their delicacy-of-construction model allows for the presence of an engaged and potentially omnipotent God, incalculable but still causally connected to the material universe, "a Supreme Intelligence [that] without interfering with the ordinary laws of matter, pervades the universe, exercising a directive energy," not unlike that which enables living beings to make choices (327). Neither religion nor morality is ever far from their thoughts, "and as in the social world a man may degrade his energy, so also in the physical world may energy be degraded" (322). Such dissipation is at once physical and moral, literal and figurative. Indeed, it is almost Dickensian.

And so we come full circle—to that sensational scene in *Bleak House* wherein Mr. Guppy and his friend Mr. Jobling come upon the charred remains of the recently exploded Mr. Krook. He has (the narrative insists, against the protestations of authorities both real and fictional) died of "Spontaneous Combustion, and none other of all the deaths that can be died" (chap. 32, 519). Suddenly, men of science and philosophy and medicine, capable of "learned talk about inflammable gases and phosphuretted hydrogen," find themselves the target of Dickens's irrepressible wit, as the wisest among them are depicted as alive with "indignation that the deceased had no business to die in the alleged manner" (chap. 33, 532). In spite of all the authorities and case studies attesting to the possibility of spontaneous combustion, to which Dickens now alludes in his novel, his fictional observers "still . . . regard the late Mr. Krook's obstinacy in going out of the world by any such by-way as wholly unjustifiable and personally offensive" (33, 532).

But for all that he dubs the combustion in question "spontaneous," Dickens suggests that while Krook's death may be incalculable, it does not follow that it is without cause. Indeed, even the offended authorities who ascribe Krook's death to obstinacy thereby posit a cause. But Dickens is more transparent about it: "The death of all lord chancellors in all courts and of all authorities in all places under all names soever, where false pretences are made, and where injustice is done . . . the same death eternally—inborn, inbred, engendered in the corrupted humours of the vicious body itself" (chap. 32, 519). His allusion to the body's "humours" suggest that Dickens is fully aware of his stubborn persistence in what will most certainly be taken as archaic science. At the same time, and true to the spirit of continuity, this gesture backward is also a gesture forward. Dickens's "humours" operate very much like Stewart and Lockyer's "directive energy": ubiquitous and speculative, they nonetheless link nonlinear effects to imperceptible causes, bringing to the fore what we might term the "romantic side" of such science writers as Stewart and Lockyer. This is both comforting and unnerving. Certainly, there is reassurance in the moral predictability inherent in Dickens's model—the certainty that the same fate will overtake all such corrupt chancellors—especially where we can see no readily discernible physical cause. There is a comforting continuity in Dickens's portrayal of this particular sensitivity to initial conditions, not unlike that ascribed to the conservation of energy. Indeed, the first-law closure that permeates the very structure of novelistic form suggests similarly that everything comes from within. And yet Dickens goes beyond the comforting assurances of Stewart and Lockyer's model. Where their model of nonlinear effects reassures us of our own free will and of the possibility of a superior intelligence, Dickens's explosive morality disturbs us in unanticipated ways. Where theirs spins a social model that remains relatively simple, predictable, and (politically as well as energetically) conservative, Dickens entangles us in a matrix of complex and nearly untraceable, networked relations.

For Stewart and Lockyer, the macroscopic history of energy accounts for the high social position and the low. In a universe neither unpredictable nor unfair, they posit that the high position of one family has been earned through the expenditure of great personal energy on the part of its founder. And even where nonlinear effects come into play, they still shore up the rightness of British expansion. The very delicacy of life itself, they claim, accounts for the marvelous advance of civilization, technology, empire, as "from an exceedingly small primordial impulse great and visible results are produced":

> In the mysterious brain chamber of the solitary student we conceive some obscure transmutation of energy. Light is, however, thrown upon one of the laws of nature; the transcendent power of steam as a motive agent has, let us imagine, been grasped by the human mind. Presently the scene widens, and as we proceed, a solitary engine is seen to be performing, and in a laborious way converting heat into work; we proceed further and further until the prospect expands into a scene of glorious triumph, and the imperceptible streamlet of thought that rose so obscurely has swelled into a mighty river, on which all the projects of humanity are embarked. (326)

Stewart and Lockyer seem as blithely unperturbed by this imperialist progression as they are by the promiscuous slippage between literal and figurative. The spark of implicitly British genius works through the progress of decidedly British technology to bring light into the heart of darkness, all fully consistent with (and seemingly predestined by) the laws of thermodynamics.

But *Bleak House* is decidedly less reassuring. After thirty-one chapters of narration, driven back and forth like the pistons of a pumping engine by the systematic alternation between the third-person omniscience of an unknown narrator and the first-person account of Esther Summerson, the whole thing explodes into an undifferentiated "we." Though we are told of Guppy and Jobling that "*they* advance slowly" toward the scene of Krook's demise (italics mine), the "they" vanishes in the ensuing confusion:

> Here is a small burnt patch of flooring; here is the tinder from a little bundle of burnt paper, but not so light as usual, seeming to be steeped in something; and here is—is it the cinder of a small charred and broken log of wood sprinkled with white ashes, or is it coal? Oh,

> horror, he IS here! And this from which we run away, striking out the light and overturning one another into the street, is all that represents him. Help, help, help! Come into this house for heaven's sake! Plenty will come in, but none can help. . . . Call the death by any name Your Highness will, attribute it to whom you will, or say it might have been prevented how you will, it is the same death eternally—inborn, inbred, engendered in the corrupted humours of the vicious body itself, and that only—Spontaneous Combustion, and none other of all the deaths that can be died. (Chap. 32, 519)

Observations, which may or may not be theirs ("here is a small burnt patch . . . here is the tinder"), rapidly give way to unknowns (is it cinder? coal?). A horror centered in being ("he IS here") panics not them, but us. And a series of insights increasingly less likely to come from anyone on the scene culminates with a consolation as universal as it gets, addressed to a mysterious "Your Highness." Even representation seems to come from nowhere and everywhere, as "We run away . . . from all that represents him." Without semblance of narrative objectivity or expectation of readerly detachment, without even a clear perspective or identifiable voice, distinctions between characters, readers, speakers, writers, and texts collapse in an unascribed "Help! Help! Help!" in response to which no help can be forthcoming. Krook's ashy remains, hardly distinguishable from a smoldering piece of wood or coal, rocket us into the street, panic-stricken and stumbling over one another as we run from a scientific anomaly that can—apparently without cause—reduce us to cinders.

This sudden and disconcerting involvement of the reader draws us into an immediate experience of chaos. It is, however, the immediacy of the experience, rather than its nonlinearity, that distinguishes it from the broader chaotic universe of *Bleak House*. Disproportionate effects abound. For example, the establishment of coffee plantations on the shores of the Niger in Borrioboola-Gha leave little Peepy with his head caught in the area railings, while a bit of distinctive handwriting amid a pile of legal papers brings down the great house of Sir Leicester Dedlock. And even the constable might think twice about his habitual refrain of "Move on," if he knew that it would eventually carry smallpox right into the heart of the story. Thus, for all that he is a proponent of proportion, Dickens brings into sharp relief how the small and the large, the near and the far, are tightly bound—how things can get very bad very quickly and in wholly unanticipated ways. And though thermodynamics undoubtedly allows Dickens, among others, "to contain the world's seeming disorder, its disjunctive elements . . . within a single system," such disjunctive elements are no less a part of the "sustained continuities linking past and present."[30] Small and large, finite and infinite, known and unknown, the familiar and the romantic, the ordinary and extraordinary are inexorably and scientifically, if incalculably, linked. Explosive effects abound in a chaos that, however unnerving, is still only natural.

Notes

1. Charles Dickens, preface to *Bleak House*, ed. Nicola Bradbury (New York: Penguin, 1996), 5. Page and chapter numbers for subsequent citations will be given parenthetically in the text.

2. For an account of this critical discussion, see Brooke Taylor, "Spontaneous Combustion: When 'Fact' Confirms Feeling in 'Bleak House,'" *Dickens Quarterly* 27, no. 3 (2010): 171–84.

3. Ibid., 172.

4. G. H. Lewes, *Leader* 4 (January 15, 1853): 64, quoted in Gordon S. Haight, "Dickens and Lewis on Spontaneous Combustion," *Nineteenth Century Fiction* 10, no. 1 (1955): 53–63.

5. Ibid., 55.

6. Barri J. Gold, *ThermoPoetics: Energy in Victorian Literature and Science* (Cambridge: MIT Press, 2010), 189–92.

7. Ann Wilkinson, "From Faraday to Judgment Day," *ELH* 34, no. 2 (1967): 225–47. Ann Y. Wilkinson argues throughout her essay to establish the validity of reading Dickens in the context of physical science, identifying the scientific underpinnings of Krook's combustion in Dickens's connection to the highly esteemed chemist and physicist Michael Faraday.

8. Crosbie Smith, *The Science of Energy: A Cultural History of Energy Physics in Victorian Britain* (Chicago: University of Chicago Press, 1998). In the chapter "North Britain *versus* Metropolis" in *The Science of Energy*, Crosbie Smith describes these competing schools of the developing science of energy, especially as regards their relations to religion and scientific naturalism, as well as contention over scientific authority.

9. George Levine, *Darwin and the Novelists: Patterns of Science in Victorian Fiction* (Cambridge: Harvard University Press, 1988), 135.

10. This phrase is taken from a description of strange-attractor chaos theory, in N. Katherine Hayles, *Chaos Bound: Orderly Disorder in Contemporary Literature and Science* (Ithaca, NY: Cornell University Press, 1990), 9–10.

11. Barri J. Gold, "The Consolation of Physics: Tennyson's Thermodynamic Solution," *PMLA* 117, no. 3 (2002): 449–64; Alfred, Lord Tennyson, *In Memoriam*, in *Alfred Tennyson: The Major Works including "The Princess", "In Memoriam", and "Maude"*, ed. Adam Roberts (Oxford: Oxford University Press, 2009), LV, 235. Stanza and page numbers for subsequent citations will be given parenthetically in the text.

12. James Prescott Joule, *The Scientific Papers* (London: Taylor and Francis, 1884), 266–67.

13. Oscar Wilde, *The Picture of Dorian Gray*, ed. Joseph Bristow (Oxford: Oxford University Press, 2008), 112.

14. Herbert Spencer, *First Principles* (Honolulu: University Press of the Pacific, 2002), 474.

15. Hayles, *Chaos Bound*, 25.

16. Tom Stoppard, *Arcadia* (Boston: Faber and Faber, 1993), 65.

17. "The Tree of Knowledge," *Dublin University Magazine* 256, no. 51 (1853): 663–75. My thanks to Stella Pratt-Smith for bringing this story to my attention.

18. William Thomson, "On the Age of the Sun's Heat," *MacMillan's Magazine* 5 (March 5, 1862): 391–92.

19. Balfour Stewart and Norman Lockyer, "The Sun as a Type of the Material Universe," part 2, *Macmillan's Magazine* 18 (1868): 319–27. Subsequent references to this article will appear parenthetically in the text.

20. Allen MacDuffie, "Joseph Conrad's Geographies of Energy," *ELH* 76, no. 1 (2009): 75–98. MacDuffie has discussed how thermodynamic principles have been misapplied in the rhetoric of European energy and efficiency—a rhetoric undermined by the actual physical pressures of work and waste (95). I have explored the social and political implications of Stewart and Lockyer's analogy in *ThermoPoetics*.

21. Balfour Stewart and P. G. Tait, *The Unseen Universe or Physical Speculations on a Future State*, 9th ed. (London: Macmillan, 1890; repr.,Whitefish, MT: Kessinger Publishing, 2003), xv. Roman numerals identify pages in the Kessinger reprint; I use these when referring to preliminary materials, such as the prefaces to the various editions. Where I use Arabic numerals, these refer to the "Article" numbers in Stewart and Tait's original text.

22. Levine, *Darwin and the Novelists*, 134.

23. Gottfried Wilhelm Leibniz, *Sämtliche Schriften und Briefe*, ed. Deutsche Akademie der Wissenschaften (Berlin: Akademie Verlag, 1923–), VI.vi.56, quoted in Larry M. Jorgenson, "The Principle of Continuity and Leibniz's Theory of Consciousness," *Journal of the History of Philosophy* 47, no. 2 (2009): 223–48.

24. Gottfried Wilhelm Leibniz, "Letter to Varignon, with a Note on the 'Justification of the Infinitesimal Calculus by That of Ordinary Algebra," in *Philosophical Papers and Letters*, 2nd ed., trans. and ed. Leroy E. Loemker (Norwell: Kluwer Academic Press, 1989), 544.

25. Gottfried Wilhelm Leibniz, "The Principle of Least Action," in *Philosophy of Science: A Historical Anthology*, ed. Timothy McGrew, Marc Alsprector-Kelly, and Fritz Allhoff (Chichester: Wiley-Blackwell, 2009), 214.

26. Wendy Wheeler, "Postscript on Biosemiotics: Reading beyond Words—And Ecocriticism," *New Formations* 64 (Spring 2008): 138.

27. Stewart and Lockyer, *Macmillan's Magazine*, 319–27.

28. This image of a gun with a delicate trigger, and much of the subsequent reasoning regarding "delicacy of construction," is reiterated in Balfour Stewart's *The Conservation of Energy* and quoted at length in *The Unseen Universe*, where the authors emphasize the "incalculability" of such systems. Tina Young Choi reads in novelistic form itself, a "Thermodynamic sensibility" that suggests a reassuring systemic self-sufficiency attached to narrative closure, which itself resists the ever-present threat of entropy. Tina Yong Choi, "Forms of Closure: The First Law of Thermodynamics and Victorian Narrative," *ELH* 74, no. 2 (2007): 301–22.

29. Stewart and Lockyer, "Sun as a Type," 324.

30. Choi, "Forms of Closure," 308–17.

CHAPTER 10

The Victorian Occult Atom

Annie Besant and Clairvoyant Atomic Research

Sumangala Bhattacharya

In August 1895, in the pastoral setting of Box Hill, Surrey, Annie Besant (1847–1933)—controversial Victorian feminist, freethinker, neo-Malthusian, Fabian socialist, anti-imperialist, and international leader of the Theosophical Society—conducted a series of experiments to determine the structure of the atom.[1] The experiments were conducted under the clairvoyant guidance of her fellow theosophist Charles Webster Leadbeater (1847–1932), a man known in his circles for the cultivation of occult powers. No laboratory setup or equipment was involved; instead, the experimenters used a form of clairvoyant meditation inspired by the yogic traditions of India to gaze into pure samples of their targeted substances. The initial results were published as an essay entitled "Occult Chemistry" in *Lucifer*, a journal published by the Theosophical Society, and later reprinted as a pamphlet in 1905. Besant and Leadbeater continued their clairvoyant research into the atom sporadically until the end of their lives, and subsequent researches resulted in three illustrated editions of the pamphlet, published by the Theosophical Publishing House in 1908, 1919, and 1951.[2]

This essay argues that *Occult Chemistry* (henceforth referred to as *OC*) should be read as an instantiation of what postcolonial theorist Ashis Nandy calls "a repertoire of the dissenting movements of science" that challenge the post-Enlightenment scientific values of rationality and objectivity.[3] The failure of *OC* with respect to the cultural power of mainstream or modern science makes it valuable as a critique of the *politics* of scientific authority.[4] Readily refutable by quantum theory and particle physics, *OC* nonetheless continues to enjoy an audience among readers who view the findings as prescient of later developments in atomic theory, such as isotopes and various exotic subatomic particles, or who value the experimenters' speculative approach as befitting the strangeness of quantum theory.[5] However, the cultural significance of the work as a critique of scientific authority has received scant attention. In his study of the trope of alchemy in the historical development of atomic theory, Mark Morrisson refers to *OC* as notable mainly for its attempt to compensate for the technological lag of instrumentation, which "offered believers a vision of a spiritual experience that was also a sensory experience of a material world," but which "simply cannot match the rhetorical power of the modern laboratory."[6] However, Morrisson does not engage with the gender and colonial implications of Besant's involvement in the experiments, which raise questions about the bases and biases inherent in how the "rhetorical power of the modern laboratory" is culturally constructed.

The lingering half-life of *OC* owes a great deal to the continued global resonances of Besant's eventful life, which greatly overshadowed the accomplishments of her fellow experimenter.[7] *OC* epitomizes a "dissenting science" perspective by blurring the distinction between experimenter and experiment: the experimenter's embodied participation is the instrumentation generating results. The contestation of scientific objectivity becomes even more apparent with the recognition that the "occult body" that performed the experiments is also a gendered body imbued with a subjectivity shaped by specific life experiences. Besant's involvement with *OC* tracks her personal journey from a crisis of religious faith to full faith in the rationality of modern science to the recognition of its limitations and a quest for alternative approaches to truth.[8] Emerging from a subjective and embodied perspective, the scientific narrative of *OC* offers a type of "situated knowledge" that presents a compelling resistance to the rationalized and progressive historiography of atomic science and quantum theory, and thus to the hegemonic cultural authority of science.[9]

The Discontents of Modern Science

Similar to many of her fellow theosophists, Besant believed that mainstream or modern science should not have a monopoly on delineating reality, a perspective informed by her experience as a crusader for women's rights and by her study of Eastern philosophies.[10] Besant understood that being a radical woman precluded her from pursuing mainstream scientific work. The small handful of Victorian female scientists who were accepted by the scientific community were genteel middle-class women who conducted themselves with great discretion in order to maintain a modicum of credibility within the mainstream scientific community. Unlike these respectable women of science, Besant had garnered considerable notoriety for her gender and class activism. When Besant legally separated from her husband, the Reverend Frank Besant, after six years of an unhappy marriage and an agonizing crisis of faith, she was given custody of their daughter (but no financial support) while her husband retained custody of their son. To support herself and her daughter, Besant moved to London, where she found a niche among the freethinkers and socialists as a writer and a powerful orator on union issues and women's rights. Besant was catapulted into national prominence by her involvement in two sensational trials: an obscenity charge for distributing birth control literature, and, immediately following that, the custody trial for her daughter.[11]

Although mainstream science, with its increasing emphasis on laboratory-intensive experimentation, was closed to Besant because of her notoriety, Eastern thought offered an alternative route to knowledge production. In 1893, Besant had succeeded the late Madame Blavatsky as leader of the Theosophical Society and moved her permanent residence to the Society's compound in Madras.[12] Besant's study of arcane Hindu philosophies in India contributed to her conviction that Indian occult practices could be harnessed in ways that intersected with, and even transcended, the work of conventional Western science. Outside of the laboratories and beyond the control of the Victorian scientific establishment, Besant hoped to show that foundational truths about life and existence were discoverable by intuitive "lay" persons and that these alternative paths to truth would reveal an essential wholeness and harmony in the universe, thereby confirming the theosophical belief that "there is no such thing as 'dead' matter; all matter is living, the tiniest particles are lives."[13]

Clairvoyant atomic research challenged mainstream scientific practices on multiple levels. First, the experimenters embraced a nontechnological methodology that repudiated the gatekeeping role of the mainstream scientific establishment in controlling the access of women and non-Europeans to the laboratory. While the complex of laboratories and equipment that had become indispensable to the production of mainstream scientific knowledge was barred to the experimenters, they contended instead that the technological complex posed an obstacle to direct human access to cosmic truths. Laboratory equipment could only show what it was designed to detect, but the human mind is not subject to such a priori limitations in what it could apprehend. Second, the premise of a holistic universe in which everything, living organisms as well as "dead matter," was imbued with a metaphysical life force was contradictory to the classificatory imperatives of Victorian science.[14] Finally, by incorporating Eastern mysticism in research on the cutting-edge Western topic of atomic structure, the experimenters contested the boundary held inviolate in modern science between consciousness and the material world. Clairvoyant atomic research mounted an important critique of the core values of rationality and objectivity that undergirded the role of modern science in authorizing colonial modernity.[15]

Twentieth-century science studies, inflected by feminist and post-colonial theories, has been increasingly attentive to the fault lines that animated Besant and her fellow discontents of mainstream science. Modern science promised a better world through the mastery of nature achieved by the disciplined exercise of scientific core values of rationality and objectivity. This project entailed the dominance and exploitation of persons or entities perceived to lack these values. As Nandy argues, this anthropocentric vision of the future, built on an overinvestment in rationality and objectivity, promotes "the dangerous human ability to separate ideas from feelings and to pursue ideas without being burdened by feelings."[16] In a similar vein, Donna Haraway deconstructs the post-Enlightenment ideal of objectivity by invoking the figure of the "modest witness," a self-effacing, neutral observer who supposedly produces representations of nature undistorted by subjective biases within "a culture of no culture," but who thereby facilitates the coexistence of the "dazzling promise" of modern science with "nastier and nastier technoscience dominations."[17] Gyan Prakash argues that the "civilizing mission" in British India, which proceeded by substituting traditional and "superstitious" beliefs with universal reason, "served as a tool for setting up the mastery of those who possessed an instrumentalist knowledge of nature over those who did not."[18] In opposition to these patriarchal and colonial agendas, clairvoyant atomic research presents an earnest contestation of the cultural authority of modern science.

The interrogations of scientific authority that culminated in *OC* were catalyzed by Besant's failure to complete her college degree and by her involvement in the Victorian cultural debate over vivisection.[19] When the

University of London began granting degrees to women in 1878, Besant immediately made plans to enroll. She passed her matriculation examinations in five subjects in 1879, winning a first class in botany and animal physiology, but failed the chemistry examination thrice. She left the university without a degree, convinced that she would never overcome the stated bias of one examiner against her notoriety.[20] The experience highlighted for her how the scientific establishment used examinations and access to laboratory resources as gatekeeping functions to exclude women and other outsiders. Similarly, Besant's engagement in the antivivisection debate led her to a powerful critique of scientific practice. Strongly influenced by the drumroll of medical discoveries using animal experimentation, including vaccination and sterilization, Besant published a modulated defense of vivisection in the name of "the freedom of science, benefactor of all that lives."[21] However, upon conversion to theosophy, Besant remorsefully and vigorously repudiated the underlying assumption she had touted earlier: that scientific knowledge justified cruelty toward creatures deemed inferior. These experiences developed her perception that the empirical and positivist thrust of modern science fragmented the natural world in its quest for knowledge.

The Fin de Siècle Atom

By the late nineteenth century, the notion of the atom as the building block of matter had become a familiar hypothesis undergirding the empiricist and positivist science of the period. In 1803, John Dalton (1766–1844) postulated that all matter was composed fundamentally of indivisible and indestructible atoms, too small to be visible with available instruments, each with a constant mass and figure.[22] Similar to Dalton's atomic hypothesis, the celebrated discovery of the laws of thermodynamics supported the Newtonian worldview in which the chaotic physical phenomena of everyday life could be rendered predictable by translation into mathematical equations dealing with mass and force. Some puzzling aspects of electricity and light were handily resolved by the postulation of ether, an all-pervasive, frictionless, surrounding medium for the propagation of light and energy that baffled all scientific efforts at observation (eventually, Einstein's theory of special relativity obviated the need for this construct).[23]

Although ether seemed a mysterious, perhaps even mystical, substance, Victorian positivist science had been able to count on the Daltonian atom as a relatively worry-free concept. However, the 1890s were heady days for the discovery of new rays and elements, with each new discovery posing quandaries about the nature of matter. For instance, in 1895, Wilhelm Conrad Röntgen (1845–1923) discovered what he called X-rays: cathode rays that could not be deflected by magnetism and that could penetrate objects opaque to light, including human flesh, thereby giving Victorians eerie glimpses of the skeletal structure of living individuals. In 1897, English physicist J. J. Thomson (1856–1940) discovered the first subatomic particle, the negatively charged electron. Thomson proposed a squishy "plum pudding" model of the atom in which rapidly orbiting electrons were suspended, like currants in a plum pudding, in a surrounding spherical mess of positive charge that held everything in place. Groundbreaking work on radioactive decay by Marie Curie (1867–1934) and Pierre Curie (1859–1906) followed in 1898. In the same year, Ernest Rutherford (1871–1937) distinguished between two types of radioactive rays, and a few years later offered the astounding theory that radioactivity causes chemical "transmutation" in which new types of matter are produced.

The fin de siècle atom was an enigma, or as Victorian man of science William Crookes (1832–1919) called it, "a tormenting mystery."[24] The Daltonian atom had given way to a strange new world of phenomena that eluded scientists' best efforts at empirical classification. This new world and the secrets it held about the structure of matter remained imperceptible to scientists as theoretical speculation outpaced advancements in the instruments of detection. Lacking the instrumentation needed to visually observe the atom, Victorian scientists could only speculate about the residual tracks and traces glimpsed in cloud chambers and spectroscopes. This empirical impasse made atomic structure a ripe and fertile field for Besant and Leadbeater's clairvoyant experiments.

The Cosmic Atom

In the opening paragraph of *OC*, Besant announces that the experiments had picked up the quest for the atom at the frontier beyond which mainstream science could not go, remedying the scientific failure of instrumentation by the employment of "astral vision," an additional human sense beyond the common five senses: "The keener and more delicate senses of the astral body are latent for the most part. . . . Yet they afford instruments for observation on the higher levels of the physical plane, and bring under direct ken objects which from their minuteness or subtlety escape ordinary vision" (211–16). This explanation positions the clairvoyant atomic researcher as a superior kind of scientist. Besant's use of the noun "ken," an outmoded term that conflates vision and recognition, suggests that the clairvoyant researchers are uniquely qualified for atomic research since their spiritual knowledge enables them to understand the phenomena observable on the "higher levels of the physical plane." Additionally, the suggestion that ordinary modes of perception may be foiled not only by the microscopic size but also by the "subtlety" of the atomic world implies that the knowledge gained by the clairvoyant researcher is always potentially outside the scope—in terms of both disciplinarity and instrumentation—of the empirical scientist.

Morrisson rightly observes that Besant uses clairvoyant atomic research to stake a theosophical claim on scientific knowledge by beating out the conventional scientists.[25] However, Besant is also positing an alternative practice for producing knowledge that parallels and occasionally intersects with mainstream science, not merely attempting to gain a foothold on the peripheries of the mainstream science. For Besant and Leadbeater, studying the atom is not an end in itself, but a path to acquiring cosmic knowledge. The experimental process begins with "an atom of a gas, and breaking it up time after time, until what proved to be the ultimate physical atom was reached, the breaking up of this last resulting in the production of astral, and no longer of physical, matter" (*OC* 217). Attaining their immediate objective of the "ultimate physical atom" (or UPA) only points the way to new terrains of knowledge about a different reality underlying physical matter.

OC is grounded in the theosophical idea of the world as a continuum of material and spiritual planes. In *Ancient Wisdom,* intended as a simplified guide to theosophical beliefs, Besant explains that ether represents a more rarefied condition of matter beyond the gaseous. There are four stages of ether, and beyond those lies the "next plane" of nature. There are altogether seven planes of existence that lead back to a singular divine will (or "Logos"). While the experiments described in *OC* involve only physical matter, the lowest and least rarefied plane, they nevertheless offer a glimpse of the higher planes, since what conventional senses perceive as the physical world is composed of "spirit-matter," a condition representing an innate potential that reaches back through all the planes to the divine will.[26]

The underlying narrative of *OC* thus presents a physical world that is permeated by the divine will as a structural component of matter. In

Ancient Wisdom, Besant explains that within the circumscribed limits self-imposed by the divine will, "the universe is born, is evolved and dies . . . its forces and energies are currents of His life; He is immanent in every atom."[27] Despite her use of the male pronoun, Besant posits divinity *not* as a gendered human-like being, but rather as a function of forces and energies bringing the universe into being and keeping it constantly in a state of flux intended to promote the evolution of all forms and matter toward higher planes. The physical world as we know it is created through manipulations of energy, resulting in the basic unit of spirit-matter, the ultimate physical atom: "The energy of the Logos . . . 'digs holes in space' in this root of matter, and this vortex of life . . . is the primary atom."[28] While conventional scientists who study the atom merely seek to extend their knowledge of nature, the clairvoyant chemist realizes that to gaze into the atom is to gaze into the infinite.

The Enigmatic Atom

Although Besant maintains a restrained scientific tone throughout *OC*, she describes an exciting and dynamic subatomic world of vibrant and continuous motion, strange forces, and puzzling formations. Its denizens inhabit an alien reality virtually impossible to convey to those who have not experienced it. Furthermore, every description or explanation implies the existence of even stranger and more esoteric realms of knowledge beyond the apprehension of mainstream science. While conventional science postulated that ether was a continuous fluid medium without molecular structure, the experimenters could see that ether was comprised of particles: "To astral vision, ether is a visible thing, and is seen permeating all substances and encircling every particle" (*OC* 217). Since ether not only *surrounds* the particles comprising matter but also *permeates* all substances, it transcends the realm of matter. This demonstrates that matter itself is an illusion created by energy and forces. What appears as a solid body is a system "composed of a vast number of particles suspended in ether, each vibrating backwards and forwards in a particular field at a high rate of velocity" (*OC* 217).[29]

Besant describes the UPA in terms that convey immense energy, unceasing movement, and enigma. A heart-shaped vortex of forces that inhabits the borderland between the material and astral planes, the UPA appears as a whole body at first glance, but is comprised of interlocking spirals of forces. Each spiral is further comprised of "spirillae, and these again of minuter spirillae" until it transcends the material plane of existence altogether (*OC* 219). Mysterious internal processes generate brilliant colors: "Sometimes one, sometimes another, is thrown into more energetic action, and with the change of activity from one spiral to another the colour changes" (*OC* 219). Although unable to pinpoint the source of the flux, the experimenters find that the vortex form is sustained and invigorated by a constant circulation of force that "pours into the heart-shaped depression at the top of the atom, and issues from the point, and is changed in character by its passage; further, force rushes through every spiral and every spirilla" (*OC* 219). The verbs "pours" and "rushes" convey an impression of power and turbulence. When these atoms combine to form elements, "Every combination begins by a welling up of force at a centre, which is to form the centre of the combination" (*OC* 219).[30]

Elements of different shapes and sizes are formed by UPAs coming together in various arrangements. In Besant's description, the spatial and relational aspects of each element also suggest personality and emotion. Hydrogen, an egg-shaped element that comprises eighteen UPAs, is in a constant state of motion that suggests nervous excitability: "It rotated with great rapidity on its own axis, vibrating at the same time, and the internal bodies performed similar gyrations. The whole atom spins and quivers, and has to be steadied before exact observation is possible" (*OC* 218). In contrast to this high-strung yet relatively simple element, oxygen is a "far more complicated and puzzling body," comprising 290 UPAs and characterized by "extraordinary activity" and "dazzling brilliancy" (*OC* 218).[31] The "comparatively quiet" nitrogen, comprising 261 UPAs, is a sedate yet intricate element that includes a balloon-like body, an egg-shaped body, and six other smaller bodies. Unlike the energetic arrangements of hydrogen and oxygen, nitrogen's arrangement is characterized by a decorous erotic restraint: "Both the balloon and egg found themselves, apparently, with the removal of their propinquity, as though they had exercised over each other some attractive influence" (*OC* 219). Curiously, the numbers of UPAs in oxygen and nitrogen, when expressed as multiples of the number in hydrogen, roughly correspond with the atomic weights of these elements as established by conventional chemistry.[32]

The experimenters' clairvoyant journey into the subatomic world is an expedition of discovery and liberation. The experimenters impact the subatomic world in significant ways through the process of observation. While the elements and their components are generally enclosed,

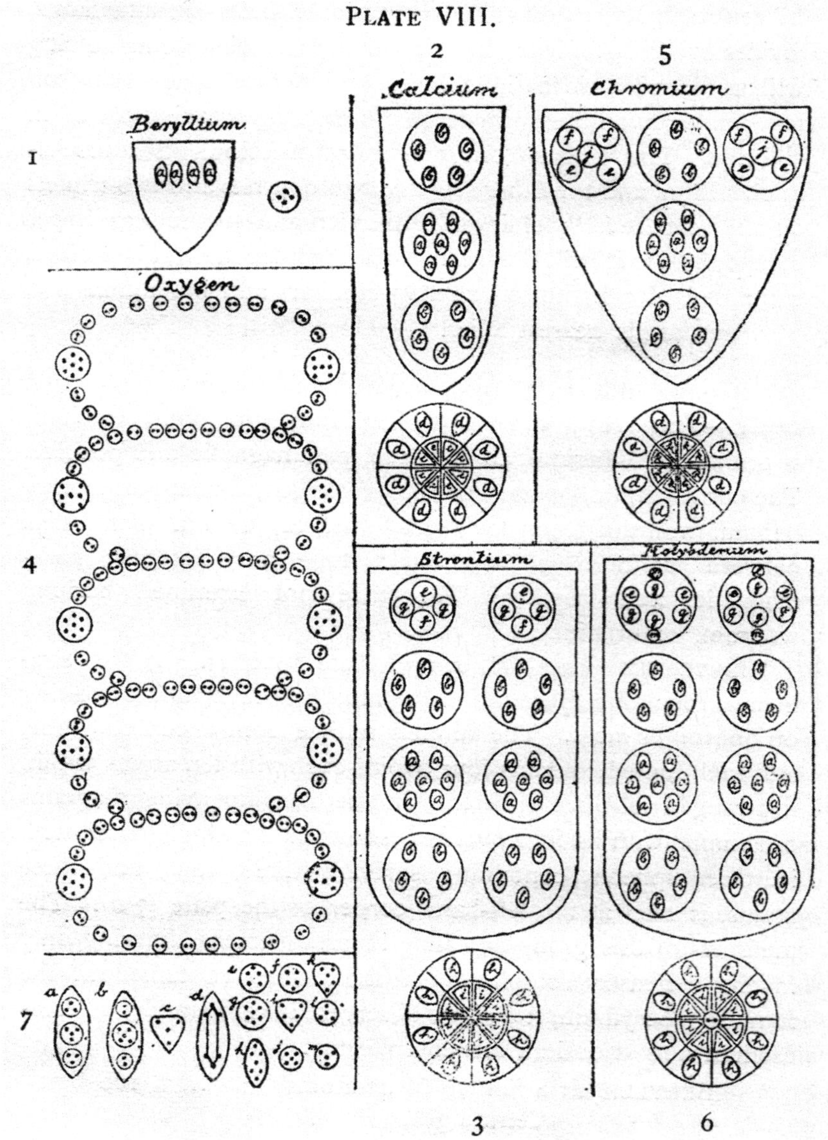

Fig. 10.1. Plate VIII, diagram of various elements: 1. Beryllium 2. Calcium 3. Strontium 4. Oxygen 5. Chromium 6. Molybdenum 7. Atomic Structures. From Annie Besant and Charles Webster Leadbeater, *Occult Chemistry: Clairvoyant Observations on the Chemical Elements* (London: Theosophical Publishing House, 1919). (Courtesy of University of Michigan and Carroll Libraries.)

the "walls" encircling them dissolve readily at the approach of the experimenter's magnified consciousness. Since these "walls" limit the element to a lower plane, the experimenters' curiosity liberates the enclosed UPAs for the move from the material to the astral plane of existence: "The falling away of the walls sets the contained atoms free, and we reach the ultimate physical atom. . . . The disintegration of this sets free particles of astral matter" (*OC* 217). Observation is thus not the prerogative of the neutral and objective scientist, but an active engagement with the world of the observed.

Magnifying Consciousness

Since the atom is beyond the reach of conventional human senses, its study demands a research methodology that bypasses the hierarchical distinction between observer and observed. Besant explains that human beings are surrounded by a world of stimuli that are beyond their sensory capabilities "because of the inability of our physical vehicle to receive and vibrate in accord with them."[33] Perhaps thinking of nineteenth-century discoveries of ultraviolet and infrared spectra, Besant adds: "Unimagined beauties, exquisite sounds, delicate subtleties, touch the walls of our prison-house and pass on unheeded."[34] However, a trained clairvoyant can open herself up to these stimuli and experiences. While scientific objectivity in mainstream science requires the separation of mind and body, clairvoyant atomic research requires that mind and body work together. The embodied and subjective knowledge produced by the clairvoyant research process highlights that what is observed is a function of the process of observation and of the observer's consciousness. As Leadbeater explains (Besant does not enter into detailed explanations of clairvoyance), clairvoyance is latent in everyone but developed enough for the apprehension of such phenomena only in a few persons.[35] The atomic experiments require an advanced technique that can be attained only by intensive training in arcane yogic practices designed to achieve *siddhi*, a transcendent state of awareness. Leadbeater initially describes this technique as conferring the power of microscopic vision, "magnifying at will the minutest physical or astral particle to any desired size, as though by a microscope," which reveals the structure of matter as "visible and living realities."[36] However, feeling that the analogy is inadequate, he eventually falls back on an orientalizing mystification (the work is marked throughout by Victorian ethnocentrism, demonstrating that even an expanded consciousness cannot always transcend cultural biases). According to ancient Oriental (Hindu) texts consulted by Leadbeater, clairvoyant magnification is "the power of making oneself large or small at will," but not in physical size: "The alteration in size is really in the student's consciousness, and not in anything outside of himself."[37]

Unlike a microscope, a magnified consciousness enables the observer not only to observe what lies beyond the limitations of physical sight, but also to interact with phenomena that confound ordinary vision. A microscope preserves the distinction between observer and observed, but a magnified consciousness evades such distinctions. Whereas a microscope functions as a prosthetic extension of eyesight, a magnified consciousness enables the experimenter to become part of the environment. Instead of selecting and focusing in on specific objects of study, the clairvoyant experimenter places her "magnified" consciousness at the same level as these objects. To be able to interact with objects in a microscopic world thus paradoxically requires one to expand oneself, and thus perhaps risk being transformed by the experience. Well aware that the methodology of *OC* violates the principle of scientific objectivity, Besant explicitly acknowledges that any kind of representation of what the observers encounter will be inadequate. Words cannot begin to convey the experience: "It is, of course, impossible to convey by words the clear conceptions that are gained by direct vision of the objects of study" (*OC* 217). The accompanying illustrations are mere third-party approximations drawn by other people based on the experimenters' descriptions. She thereby signals that any assessment of the verity of these experimental results requires trust in the experimenters' integrity and a lively imagination that can overcome the limitations of language. Such an approach is antithetical to mainstream scientific process. Modern science is supposed to be a universal practice, open to anyone who has the necessary skills and resources. Additionally, scientific truth claims derive from experimental verification by disinterested scientists who rely on the objectivity of instrumentation. In contrast, clairvoyant powers are sufficiently developed only in a few, rare individuals. Furthermore, a rare person possessing the powers required to repeat the experiment is unlikely to be a disinterested party, since to develop these powers, one needs training from experienced mentors. Moreover, *OC* can only proclaim that the experimenters saw what they saw and that others sufficiently trained might also see the same phenomena.

In foregrounding the failure to conform to the normative values of modern science, Besant invites an interrogation of the assumptions undergirding those values. Modern science cannot be truly universal since it requires talented scientists (not everyone has the faculty for scientific research). The extensive training process affiliates scientists with their mentors and the scientific establishment, thereby potentially tainting the claim to scientific neutrality and objectivity. Additionally, scientific instrumentation is often inaccessible to outsiders, who must therefore *trust* in the integrity of scientists and the scientific process, much as the audience of *OC* must trust the experimenters. Latour observes that in highly specialized areas of modern science, "The proof race is so expensive that only a few people, nations, institutions or professions are able to sustain it" (174).[38] The authority and credibility of modern science derive from enormous financial and physical investments, which can create points of pressure on the scientific process. In contrast, Besant and Leadbeater's do-it-yourself methodology contests the authority of truths produced by elaborate laboratory resources that are inaccessible to ordinary persons and asserts that knowledge about the world can be produced in venues and contexts other than the metropolitan Western scientific establishment.

A Dissident Science

By the 1890s, Western science had become synonymous with modernity and rationality, with momentous cultural consequences for the world. As Latour notes, the complexity of laboratories has grown to such an extent that "laboratories are now powerful enough to define reality."[39] According to Prakash, the colonial context further intensified the authority of science, as "Science came to signify not just scientific research in laboratories but also new forms of rule and authority" and "The reach of science's authority extended far beyond the laboratory to function as a grammar of social and cultural transformation."[40] Yet, as physicist David Bohm has argued, modern science presents a reality flawed by "deep and radical fragmentation, as well as thoroughgoing confusion" due to the scientific resistance to the insights of quantum theory, which point to "a deeper reality in which what prevails is unbroken wholeness."[41]

OC appropriates a cultural moment in the British Raj and the Indian nationalist movement to articulate the yearning for an "outsider" and dissident stake in knowledge production.[42] By yoking archaic and arcane Indian (Hindu) epistemologies to atomic science, *OC* seemingly anticipates Bohm's vision of a holistic reality. For the anti-imperialist and feminist Besant, the strangeness of atomic physics called for a strange science that embraced enigmatic and mystical explanations of the nature of matter. The experimenters' repudiation of elaborate and expensive laboratory setups speaks to a Romantic privileging of individual consciousness while empowering pure theoretical research in a mode that might be pursued even in the resource-starved colonial domain. Through her clairvoyant atomic research, Besant suggests that the geopolitical arc connecting the metropolitan and imperial center of London to the colonial periphery of India may be traveled in reverse, with the colonial peripheries as the source of more significant knowledge than empirical facts generated in modern laboratories. At this present time, when mainstream science deployed across the world in the service of modernization and globalization further co-opts voices and ways of life at the peripheries, *OC* reminds us that the "mystical" and "irrational" can encode dissent from and critique of the hegemonic cultural power of science that legitimizes the domination and fragmentation of nature.

Notes

I dedicate this article to my father, Sushil Kumar Bhattacharya (1922–2006), and my mother, Gitanjali Bhattacharya (1929–2013), and other elders who instilled in me a passion for strange science and other intellectual curiosities.

1. The Theosophical Society was a quasi-religious organization principally founded by Helena Petrovna Blavatsky (1831–1891), Henry Steel Olcott (1832–1907), and William Quan Judge (1851–1896). The organization combined aspects of various world religions, including Hinduism, Buddhism, Christianity, and Western esoteric thought. See Marian A. Fisher, "Theosophy," in *Victorian Britain: An Encyclopedia*, ed. Sally Mitchell (New York: Routledge, 2011), 803–4.

2. Annie Besant, "Occult Chemistry," *Lucifer* 17 (November 1895): 211–19; Besant and Charles Webster Leadbeater, *Occult Chemistry: Clairvoyant Observations on the Chemical Elements*, ed. Alfred Percy Sinnett (London: Theosophical Publishing House, 1919), http://www.gutenberg.org/files/16058/16058-h/16058-h.htm); Besant and Leadbeater, *Occult Chemistry: Investigations by Clairvoyant Magnification Into the Structure of the Atoms of the Periodic Table and Some Compounds*, ed. C. Jinarajadasa (Adyar, Madras, India: Theosophical Publishing House, 1951). Unless otherwise indicated, further citations refer to the essay version.

3. Ashis Nandy, *The Intimate Enemy: Loss and Recovery of Self under Colonialism* (New Delhi: Oxford University Press, 1983), 12.

4. I use the terms "modern science" and "mainstream science" interchangeably to describe Western post-Enlightenment empiricist scientific practices that hold objectivity and rationality as core values. For critiques of modern science, see Sandra Harding, *Sciences from Below: Feminisms, Postcolonialities, and Modernities* (Durham, NC: Duke University Press, 2008); Bruno Latour, *Science in Action: How to Follow Scientists*

and Engineers through Society (Cambridge: Harvard University Press, 1987); and Nandy, *Intimate Enemy.*

5. For an argument that the experimenters had detected quarks, see Steven M. Phillips, *The Extrasensory Perception of Quarks* (London: Theosophical Publishing House, 1980). For a scientific refutation of *OC*, see Michael J. McBride, "Serious Scientific Lessons from Direct Observations of Atoms through Clairvoyance," Department of Chemistry, Yale University, last modified December 6, 1999, accessed January 2, 2013, https://webspace.yale.edu/chem125/125/history99/8Occult/OccultAtoms.html. For the influence of *OC* on Francis Aston's discovery of neon isotopes, see Jeff Hughes, "Occultism and the Atom: the Curious Story of Isotopes," *Physics World*, September 2003, 31–35. For the lingering influence of Besant in India, see B. Sivakumar, "Direct Knowledge," *Times of India*, January 7, 2012, http://timesofindia.indiatimes.com/life-style/Direct-knowledge/articleshow/10725961.cms. See also the ironic but extensive treatment of *OC* in Arthur Greenberg, *From Alchemy to Chemistry in Picture and Story* (Hoboken, NJ: John Wiley, 2007), 599–601.

6. Mark S. Morrisson, *Modern Alchemy: Occultism and the Emergence of Atomic Theory* (Oxford: Oxford University Press, 2007), 95.

7. Although Leadbeater was an influential member of the Theosophical Society, his role was dogged by scandal and he never achieved Besant's level of global prominence. Furthermore, Leadbeater's interest in the experiments focused on the mechanism of clairvoyance, whereas Besant was interested in the scientific implications of the findings. She was active throughout her life in promoting scientific education and practices that incorporated Eastern thought.

8. Carol Hanbery MacKay argues that Besant's life can be best described as an unfolding pattern of "painful deconversions" in which Besant "uncovers the elements of self-contradiction within her current belief system . . . then she explores forbidden knowledge; and finally, she embraces a more encompassing structure—one which promises a more optimistic program of dissolving those boundaries that separate human souls." *Creative Negativity: Four Victorian Exemplars of the Female Quest* (Stanford: Stanford University Press, 2001), 102. Gauri Viswanathan sees in Besant's life the convergence of multiple strands of cultural change, including domestic dissent in Britain and anticolonial struggles abroad, and argues that Besant's progressivism was complicated by a strong belief in evolutionary theory, *Outside the Fold: Conversion, Modernity, and Belief* (Princeton, NJ: Princeton University Press, 1998), 190.

9. Donna J. Haraway uses the term "situated knowledge" for a "feminist objectivity" deriving from "a doctrine of embodied objectivity that accommodates paradoxical and critical feminist science projects." Haraway, "Situated Knowledges: The Science Question in Feminism and the Privilege of Partial Perspective," *Feminist Studies* 14, no. 3 (1988): 581.

10. For more on the status of women scientists in the nineteenth century, see Richard Holmes, "The Royal Society's Lost Women Scientists," *Guardian Observer*, November 20, 2010, http://www.theguardian.com/science/2010/nov/21/royal-society-lost-women-scientists.

11. In 1877–78, Besant and her mentor Charles Bradlaugh were sentenced to six months in prison on obscenity charges for publishing and distributing a neo-Malthusian pamphlet advocating birth control. The conviction was overturned on a technicality. A month after the final verdict, Frank Besant applied for custody of their daughter. Besant's case was the first one to be tried under the Custody of Infants Act of 1873, which allowed mothers to have access to or custody of children after separa-

tion. Besant was forced to settle for occasional visitation rights, which she eventually renounced to protect the children's emotional health. See Arthur H. Nethercot, *The First Five Lives of Annie Besant* (Chicago: University of Chicago Press, 1960), 107–44; Carol Hanbery MacKay, introduction to *Autobiographical Sketches* by Annie Besant (1885), ed. Carol Hanbery MacKay (Peterborough, ON: Broadview, 2009), 39.

12. Besant converted to theosophy almost immediately after a meeting with the charismatic Blavatsky, and quickly became a rising star of the Theosophical Society. When Blavatsky died in 1891, Besant succeeded her as the leader of the Society in Europe and India. Besant taught herself Sanskrit and conducted intensive studies of ancient Hindu texts. She also involved herself in educational and political reforms in India and became an influential, if controversial, voice in Indian nationalist politics. Nethercot, *The Last Four Lives of Annie Besant* (Chicago: University of Chicago Press, 1963), 213–91. For Besant's role in Indian nationalism and the conflict with Gandhi, see Raj Kumar, *Annie Besant's Rise to Power in Indian Politics, 1914–1917* (New Delhi: Concept Publishing, 1981).

13. Besant, *The Ancient Wisdom: An Outline of Theosophical Teachings* (1897; Adyar, Madras, India: Theosophical Publishing House, 1969), 48.

14. Besant's views appear in some respects aligned with the natural history concept of vitalism, which held that living organisms, unlike inanimate matter, were subject to an unspecified vital force that defied physical and chemical analysis. Besant goes further than most vitalists in repudiating a fundamental distinction between living organisms and "dead" matter. By the end of the nineteenth century, vitalist philosophies had been largely discredited for their reliance on a metaphysical concept instead of a genuine explanation for biological phenomena. For more on the demise of vitalism under pressure from mainstream scientific materialist approaches, see Garland E. Allen, "Mechanism, Vitalism, and Organicism in Late Nineteenth and Early Twentieth-Century Biology: The Importance of Historical Context," *Studies in History and Philosophy of Biological and Biomedical Sciences* 36 (2005): 261–83.

15. Numerous scholars have discussed the implication of Western science in sustaining the violence and inequities institutionalized in various practices and ideologies of modernity, especially in the context of colonialism and the neocolonial state. See, for instance, Abraham Itty, *The Making of the Indian Atomic Bomb: Science, Secrecy and the Postcolonial State* (London: Zed Books, 1998).

16. Nandy, *Intimate Enemy*, 278.

17. Donna Haraway, *Modest_Witness@Second_Millennium.FemaleMan©-Meets_OncoMouse™: Feminism and Technoscience* (New York: Routledge, 1997), 41.

18. Gyan Prakash, *Another Reason: Science and the Imagination of Modern India* (Princeton, NJ: Princeton University Press, 1999), 5.

19. Noted suffragist and activist for women's rights Frances Power Cobbe helped establish the National Anti-Vivisection League in 1875. The reports of a Royal Commission on vivisection (1875–76) led to the Cruelty to Animals Act of 1876. However, the act did not provide public accountability of licensing practices, and supporters of the antivivisection movement continued their agitation. See John Ranlett, "Animal Protection," in Mitchell, *Victorian Britain*, 26–27.

20. Besant encountered several other instances of prejudice in the course of her studies, e.g., she was denied access to facilities and only allowed to use them after regular hours; and her name was deliberately omitted from public rolls of successful students (Nethercot, *First Five Lives*, 179–82).

21. Annie Besant, "Vivisection," in *The Moral Aspects of Vivisection* (London: Wil-

liams and Norgate, 1882), 8, https://books.google.com/books?id=9xsdTs46pyUC&source=gbs_navlinks_s.

22. For more on Dalton's atomic theory, see Mary Jo Nye, *Before Big Science: The Pursuit of Modern Chemistry and Physics, 1800–1940* (Cambridge: Harvard University Press, 1999), 28–56.

23. For a history of atomic science in Britain, see Alex Keller, *The Infancy of Atomic Physics: Hercules in His Cradle* (Oxford: Clarendon, 1983). For energy physics in Victorian literature, see Barri Gold, *ThermoPoetics: Energy in Victorian Literature and Science* (Cambridge: MIT Press, 2010). For a layperson's introduction to quantum theory, see John Polkinghorne, *Quantum Theory: A Very Short Introduction* (Oxford: Oxford University Press, 2002).

24. See the presidential address to the British Association for the Advancement of Science by William Crookes, "Inaugural Address," *Nature* 58 (May–September 1898): 444, https://books.google.com/books?id=z4xFAAAAYAAJ&dq=Nature+September+1898&source=gbs_navlinks_s).

25. Morrisson, *Modern Alchemy*, 67–68.

26. Besant, *Ancient Wisdom*, 48.

27. Ibid., 45.

28. Ibid., 47.

29. With some important differences, Besant's description of the subatomic world as one of perpetual vibrations evokes the vortex theory of matter, originally proposed by mathematical physicist William Thomson, later Lord Kelvin (1824–1907), which held that atoms were linked and knotted vortices in ether. See Helge Kragh, "The Vortex Atom: A Victorian Theory of Everything," *Centaurus* 44 (2002): 32–114.

30. Subsequent editions significantly modified these initial descriptions. The 1919 edition, extensively edited by theosophist Alfred Percy Sinnett (1840–1921), describes the UPA as possessing two gendered cognate forms. In the 1951 deluxe edition of *OC*, significantly altered by the editor C. Jinarajadasa, the term "ultimate physical atom" is replaced by the Sanskrit term *Anu. Anu* refers to "a minute particle, usually translated as an atom, [that] is described in early [Vedic] philosophies as a material component of the universe." Roshen Dalal, *Hinduism: An Alphabetical Guide* (New Delhi: Penguin India, 2011), 33.

31. Curuppumullage Jinarajadasa, Plate VIII, illustration in *Occult Chemistry: Clairvoyant Observations on the Chemical Elements* by Annie Besant and Charles Webster Leadbeater, ed. Alfred Percy Sinnett (London: Theosophical Publishing House, 1919), 52.

32. Such moments of inexplicable apparent consistency with mainstream scientific findings likely contribute to the continuing appeal of *OC*. Greenberg, *From Alchemy*, 599–601.

33. Besant, *Ancient Wisdom*, 57.

34. Ibid.

35. Leadbeater, *Clairvoyance* (London: Theosophical Publishing Society, 1903), 3, http://www.gutenberg.org/files/29399/29399-h/29399-h.htm.

36. Ibid., 47–49.

37. Ibid.

38. Latour, *Science in Action*, 174.

39. Ibid., 93.

40. Gyan Prakash, "The Modern Nation's Return in the Archaic," *Critical Inquiry* 23, no. 3 (1997): 538.

41. David Bohm, *Wholeness and the Implicate Order* (London: Routledge, 1980), xiii–xv.

42. While many Indian nationalists argued that a modern scientific establishment controlled by Indians was crucial to the modern nation, other Indian elites, especially among Hindus, turned to Orientalist studies of ancient Sanskrit texts to counter the resource and credibility gap by claiming for India a forgotten scientific heritage that anticipated the discoveries of modern science. Prakash argues that "the authority of the *Vedas* as science and as a sign of the nation was part of a general revaluation and positioning of the Hindu past as an expression of the nation," "Modern Nation's Return," 543. Besant's work could also be seen as supporting this Orientalist strain regarding science in the Indian nationalist movement. See also Deepak Kumar, *Science and the Raj, 1857–1905* (New Delhi: Oxford University Press, 1995).

CHAPTER 11

Inductive Science, Literary Theory, and the Occult in Edward Bulwer-Lytton's "Suggestive" System

Anna Maria Jones

> If, then, some tyrannical Afrite, wroth with my modest disavowal of a system, or my arrogant pretensions to suffer my thoughts to grow without cord or stake, should say to me, "System of some kind thou shalt choose," my system should be the suggestive, because it is given to few men to prove, and all men to suggest. . . . Thought is valuable in proportion as it is generative. If vital itself, though it be but a germ, it vitalises thoughts in others which may bloom into petals, or mature into fruits not vouchsafed to the thinker in whom it originates. I cast my thoughts freely abroad; let the winds waft them loose. It is according to the soil on which they fall that they will be sterile or fertile.
>
> —Edward Bulwer-Lytton, *Caxtoniana* (1863)[1]

When one thinks of the foremost scientific thinkers of the nineteenth century, Edward Bulwer-Lytton (hereafter Bulwer) is unlikely to make the short list. Indeed, despite his many successes as a popular novelist and essayist during his lifetime, today he is even unlikely to make the list of great Victorian literary theorists or practitioners. Yet, it is worth taking Bulwer's intellectual labor seriously on both counts. Bulwer's literary theory is inextricably bound with his understanding of Baconian science, and both of these concerns are interconnected with his interest in the occult. This is not to say that Baconian induction drives the plot of Bulwer's fiction in the way, for example, that astronomy governs Thomas Hardy's *Two on a Tower* or evolutionary theory underpins H. G. Wells's *Time Machine.* Inductive science is not a fixed point by which to navigate Bulwer's oeuvre or, in any straightforward way, a "through line" that connects his works. Rather, he appropriates Bacon idiosyncratically and with considerable license in both his fiction and his essays—indeed, Bulwer's creative repurposing of Baconian science in the service of his literary and occult theories demonstrates the "suggestive system" that he outlines in the epigraph: in Bulwer's work, Bacon's scientific method "mature[s] into fruits not vouchsafed to the thinker in whom it originate[d]."[2]

As I will argue, Baconian induction is foundational to Bulwer's "suggestive" system of intellectual inquiry, which insists upon the impossibility of predicting the effects of a text on its readers and which, consequently, privileges eclectic collection of facts, texts, and ideas over some more "systematic" pursuit of knowledge. Bulwer's creative appropriation of Baconian science enables him to draw connections analogically across various scientific, occult, and literary fields of study and to blend opposing epistemological and aesthetic impulses, materialist and idealist. In this essay I trace Bulwer's "suggestive system," primarily in *Caxtoniana,* his 1863 collection of essays, which treat, as the subtitle asserts, "life, literature, and manners," and which first appeared serially in *Blackwood's Edinburgh Magazine* from February 1862 through October 1863; I end, however, with brief attention to *Zanoni* (1842), Bulwer's first occult novel, in which he stages an evocative demonstration of his inductive system of reading at work. Bulwer's insistence on the power of thought to "mature into fruits not vouchsafed to the thinker in whom it originates,"[3] I argue, decouples text from author, thereby enabling Bulwer to consider the discursive afterlives of ideas—his own and others'—as they continue to circulate and to shape readers in surprising ways and in distant futures.

Bulwer adopts his inductive methodology from William Whewell and William Godwin, both his onetime mentors.[4] In his second occult novel, *A Strange Story* (1861), for example, the narrator's mentor, Dr. Faber, quotes Bacon (via Whewell) to encourage in the skeptical narrator a faith in the divine to undergird his scientific pursuits:

> I see on your table the very volume of Bacon which contains the passage I commend to your reflection. Here it is. Listen: "Take an

> example of a dog, and mark what a generosity and courage he will put on when he finds himself maintained by a man who, to him, is instead of a God. . . . So man, when he resteth and assureth himself upon divine protection and favour, gathereth a force and faith which human nature could not obtain."[5]

In *A Strange Story,* as in *Caxtoniana,* Bulwer takes his cue from Whewell's 1833 Bridgewater Treatise, *On Astronomy and General Physics Considered with Reference to Natural Theology,* invoking Bacon to argue against the "hallucination by which Nature is left Godless—because Man is left soulless."[6] But it is William Godwin who provides Bulwer with a pattern for harnessing the Baconian method of inductive reasoning to a theory of literature and readers.

According to Godwin in *The Enquirer,* his 1797 collection of essays on literature, culture, and knowledge, inductive reasoning is crucial to understanding what books actually *do.* Whereas a book may be written to illustrate a particular moral, its real influence on readers, or "tendency," as Godwin calls it, may diverge widely. The moral, he writes, is that "ethical sentence to the illustration of which the work may most aptly be applied," but the tendency "is the actual effect it is calculated to produce upon the reader, and cannot be completely ascertained but by experiment."[7] To attend too rigidly to a book's moral—what general principle it illustrates *in theory*—is to miss what it may be doing to readers *in practice.* And, Godwin warns, "To ascertain . . . the genuine tendency of a book, is a science peculiarly abstruse."[8] But he also suggests in the preface to *The Enquirer,* with indirect allusion to Bacon, that

> the intellectual eye of man, perhaps, is formed rather for the inspection of the minute and near, than of immense and distant objects. We proceed most safely, when we enter upon each portion of our process, as it were, *de novo*; and there is danger, if we are too exclusively anxious about the consistency of system, that we may forget the perpetual attention we owe to experience, the pole-star of truth.[9]

Godwin's emphasis on de novo analysis—which is to say, starting afresh with each new object of study instead of moving too quickly to conclusions or proceeding too confidently from preconceived notions—echoes Baconian edicts: to interpret nature through the careful collection of facts and to free one's intellect from biases (*idola*).[10] Indeed, Godwin claims, "an incessant recurrence to experiment and actual observation . . . is the method adopted" in *The Enquirer* itself, rather than "laying down one or two principles and then developing them and following them to a variety of inferences."[11] In other words, Godwin undertakes an inductive method in his own literary and cultural meditations.

In *Caxtoniana* Bulwer cites Godwin explicitly in describing this dichotomy and notes:

> A writer may present to you, at the end of his book, some unexceptionable dogma which parents would cordially admit into the copybook ethics of their children, yet, in the process of arriving at his harmless aphorism, he may have led the mind as much astray into mischief as it is in his power to do. On the other hand, a writer may seek to work out a proposition, from the moral truth of which there would be a very general dissent, and yet be either harmless, or often instructive and elevating.[12]

To accommodate this unpredictability, Bulwer, like Godwin, proposes an approach to intellectual inquiry that privileges minute and varied observation and experimentation and that resists too-hasty judgments, acknowledging even the value of pseudoscientific missteps as much as, or perhaps even more than, the elucidation of infallible laws:

> It is noticeable how much even the fallacies of a great writer serve, not the less effectually, because indirectly, to the advancement of truth, by stimulating the energies of the writers who oppose the fallacies, and, in so doing, strike out new ideas and fresh discoveries. How much of his researches into alchemy may have warmed and emboldened the imagination of Newton, in whom imagination seems to have been only less powerful than reason![13]

Bulwer's invocation of Newton here is typical of much of *Caxtoniana,* which works on multiple levels: to link scientific experimentation to creativity, to encourage the suspension of bias against the occult, and to draw analogies among different kinds of intellectual endeavor. As he writes, "All the acquisitions of the human intellect are relations to each other. . . . If there be some specialty in art, literature, science, active life, in which we can best succeed, that specialty is improved and enriched by all the contributions obtainable from other departments of study."[14] And, so, the "indirect" benefit of any given idea, text, or author cannot be determined except by observation.

I follow Lawrence Poston in tracing the outlines of a "Godwinian" Bulwer. Poston reads Bulwer in "triangulation" with Godwin and Mary Shelley as political novelists who "engage issues of power and control" in psychological, occult, and supernatural terms.[15] I would certainly agree with this characterization, but I read Bulwer's project, particularly in *Caxtoniana* and in his occult fictions—*Zanoni*, "The Haunted and the Haunters" (1859), and *A Strange Story*—as more epistemological than political. And this is why I find his invocations of Baconianism (heretofore unexamined by scholars, as far as I know) of particular interest. By exploring Bulwer's "suggestive" mode, I propose a tentative explanation of the inconsistencies and contradictions in his ideas, which have often been attributed (more or less disparagingly) to a canny ability to capitalize on multiple literary markets rather than to any genuine intellectual commitments. For example, Joseph Fradin's characterization in 1961 of Bulwer as a mostly unsatisfying if also "highly symptomatic novelist, sensitive not merely to changes in the barometer of taste but to changes in the intellectual atmosphere," may appear somewhat dated now, but it is in many ways still typical.[16] Christopher Lane's treatment of Bulwer's philosophical exploration of misanthropy is considerably more flattering to him, arguing that "his essays thicken recent interest in hatred, antagonism, and anti-communitarianism—voiced by such disparate thinkers as Giorgio Agamben, Etienne Balibar, Jacques Derrida, Ernesto Laclau, Claude Lefort, Carl Schmitt, and Slavoj Žižek," but Lane, too, cites Bulwer's value as "something of a Victorian weathervane."[17] Other critics have been much less generous.[18]

Marie Mulvey-Roberts notes wryly that "Bulwer-baiting is not just a recent phenomenon, it was also a national sport for his contemporaries as well. . . . Bulwer appeared to some to be a shifting, ephemeral and nebulous figure, rather like his own literary creation, the 'Dweller of the Threshold,' who appears in *Zanoni*."[19] I suggest that this very "shifting, ephemeral and nebulous" quality bespeaks Bulwer's desire to "let [his] thoughts grow without cord or stake," according to his suggestive system. As he argues, with recourse to scientific language, "'Non fingor [*sic*] hypotheses,' said Newton, with a scorn we revere in a Newton, to whom scorn was so rare. Still, if Newton disdained an hypothesis, he rejoiced in a guess."[20] Here again, Bulwer's idiosyncratic Newton is both a "mechanical" scientist pursuing his Baconian method of experimentation and also an intuitive and creative genius, a blend of occult visionary and inductive observer, on a level with the greatest of literary geniuses: "I was mistaken in calling Shakespeare 'peerless' in the gift of clairvoyance—Newton's clairvoyance is not less marvellous than Shakespeare's. To imagine the things they have never seen, and to imagine them accurately, constitutes the poetry of philosophers, as it constitutes the philosophy of poets."[21]

Rather than reading Bulwer's enthusiasm for occult disciplines, from alchemy to mesmerism to clairvoyance, as indicative of an unfortunate lack of skepticism (an extreme case of what was surprisingly common among [otherwise serious] Victorian intellectuals like Harriet Martineau, Thomas Carlyle, and George Eliot, among others) or of a venal desire to profit from contemporaneous tastes, we might, rather, read this eclecticism and credulity in the spirit of Bulwer's own "suggestive" system: as providing "generative" value for him and his readers.[22] In this I follow Mark Knight's suggestion that "rather than consigning these esoteric studies (and Bulwer with them) to the margins of nineteenth-century culture, it is more helpful to consider the light they shed on some of the main scientific and philosophical debates of the period."[23] If Bulwer's image of a clairvoyant Newton or a magical Bacon makes for "strange science," it also offers a useful reminder that the categories into which we divide our own intellectual labors—scientific or humanistic, material or spiritual, critical or creative, empirical or intuitive—were by no means so discrete for Victorian thinkers.

Bacon, Newton, and Victorian Epistemology

The fact that emerging scientific theories and new discoveries were central to Victorian literary and cultural discourses is no new insight, but the importance of the debate over the scientific method itself has received less attention in Victorian literary studies, despite the fact that, as George Levine notes, "Bacon . . . was a pervasive presence in the consciousness of theorists of science, and his way of seeing the world was a more or less 'official' and certainly pervasive one in Victorian England."[24] Francis Bacon's inductive method—wherein the scientist must start from observable facts and work (only slowly and patiently) from these to the formulation of theories or broad principles, as opposed to working deductively from the abstract theory toward description of particulars—was hotly debated by Victorian intellectuals. His privileging of fact-gathering appealed to empiricists, and his emphasis on the practical application of scientific discoveries made him attractive to utilitarians and industrialists. Philosophical materialists pointed to Bacon's emphasis on observation as justifying the primacy of sensory phenomena; idealists sometimes criticized him as too narrowly focused on minutia, and thus blind to the need for larger ideas and abstractions, but also sometimes presented him sympathetically as a theological scientist. Jonathan Smith argues that the waning of the popularity of Baconian induction as the nineteenth century progressed was linked to the rise in a hypothetico-deductive method that "sought to portray science as an imaginative, speculative, creative enterprise." He also suggests that this decline in Bacon's fortunes was linked to a concomitant rise in Newton's popularity.[25] However, as Richard Yeo notes, and as becomes clear reading the Victorians on Bacon, even applying the terms "Baconian" and "Baconianism" to nineteenth-century science and scientists is problematic: "These terms . . . have been objects of controversy and multiple interpretation to an extent which renders them practically useless as simple descriptive epithets."[26] Certainly, Bulwer's versions of Baconian and Newtonian science bear the stamp of his creative license.

Nevertheless, Bulwer is clearly participating in the Victorian attempt to yoke imagination to scientific inquiry. Interestingly, however, for Bulwer this does not involve a rejection of Bacon's empirical method as "too sterile, too mechanical, and too impersonal to capture the artistic quality of the scientist doing science,"[27] but rather requires a conception of Bacon himself as accommodating imagination while *also* being methodologically committed to fact-gathering. For example, for Bulwer, as for his fellow believer in mesmerism Harriet Martineau—to whom he originally suggested mesmerism as a cure for her illness[28]—Bacon's adherence to unbiased observation confers legitimacy on explorations of the occult, wherein the "facts" of successful mesmerism, inexplicable clairvoyance, and other wonders had, according to their proponents, not *yet* been collected into incontrovertible knowledge or coherent systems. Martineau's 1851 collaboration with Henry George Atkinson, *Letters on the Laws of Man's Nature and Development*, which extols the virtues of atheism, Baconianism, and mesmerism in equal measure, begins with "mottoes" from Bacon, chemist Sir Humphrey Davy, and phrenologist Franz Gall (among others). Also, the book includes numerous invocations of Bacon like this one:

> Many persons, from fancying that mesmerism and *clairvoyance* indicate a spiritual existence, or something supernatural, have become converted from scepticism to the belief of a future life. . . . but mes-

> merism and *clairvoyance* are as natural as the instinct of animals, and no more wonderful. . . . We must follow our great master Bacon, and make a stand against the fallacy of natural theology.[29]

Indeed, for authors aiming to distance themselves and their inquiries from theology, their reverence for Bacon and natural law approaches religious zealotry, as G. H. Lewes pointed out in his scathing two-part review in the *Leader*.[30]

Martineau and Atkinson's *Letters* are interesting in relation to Bulwer for a couple of reasons. First, they appropriate Bacon in very similar ways to Bulwer, but from almost opposite poles of the idealism-materialism debate, thus underscoring the slipperiness (and ubiquity) of Baconianism in Victorian intellectual circles. Moreover, while Bulwer receives slightly more attention from literary critics today than Martineau,[31] they are both figures who were extremely influential and widely read in their day, but who have since fallen into relative obscurity. Our picture of the Victorians' relationship to "strange science" remains incomplete without considering these unlikely convergences as well as the "usual suspects" (e.g., Darwin, Eliot, Lewes, Spencer) in the circulation of Victorian scientific discourses.

Bulwer, though staunchly opposed to the materialism that Martineau and other Comtean positivists championed, puts Bacon to similar use, urging the suspension of prejudices in the examination of "some of the modern thaumaturgia":

> Possibly, if a philosopher who possessed in an equal degree the virtue of candour and the acuteness of science, would condescend to examine, as Bacon and Newton would unquestionably have examined [them] . . . possibly he might either make an immense progress in our knowledge of the laws of nature, or prevent incalculable mischief in the spread of a new superstition.[32]

A philosopher is never justified in denying phenomena just because they seem, on the face of things, to contravene the laws of nature, for "if a philosopher is to pronounce for himself what is impossible and what is not, there would soon be no philosophy at all."[33] Bulwer's suppositious Bacon and Newton here are together praiseworthy for their willingness to observe without bias. Disposing of his imagined naysayer, Bulwer concludes: "Certainly Newton would not have so answered, because he never refused to examine."[34]

Of course, these (mis)appropriations of Bacon and Newton may be evidence in support of Smith's claim that "the decline in Bacon's reputation as a methodologist had its roots in part in the work of those nineteenth-century thinkers who admired him."[35] G. H. Lewes's acerbic criticism, in *Comte's Philosophy of the Sciences* (1853), of those who sought to link Bacon and Newton reads uncomfortably as if he had Bulwer specifically in mind (though he might very well have been thinking of Martineau and Atkinson). Referring to the same line that Bulwer misquotes, Lewes writes: "Newton's assertion—*Hypotheses non fingo*—I make no hypothesis—has been incessantly repeated by men who fancy themselves Baconian thinkers when they restrict their incompetence to what they call 'facts.'" These "ideas of science," Lewes declares, "are utterly irrational."[36] Tone aside, Lewes's comment points to Bacon's and Newton's multipurpose utility for materialists, idealists, occultists, among others, as, indeed, the Victorians themselves were well aware.

Thus, Bulwer finds Bacon's empirical method appealing because, in his interpretation, it accommodates both spiritual faith and the pursuit of material knowledge. As "L——," a semiautobiographical persona in Bulwer's 1830 Socratic dialogue "The New Phaedo," remarks to his interlocutor, "[Bacon] makes us feel less earthly in our desires, by making us imagine ourselves *wiser*—the love of a divine knowledge inspires and exalts us. . . . With the same hand that limits our progress on earth, he points to the illimitable glories of heaven."[37] "The New Phaedo" appeared much earlier than *Caxtoniana*, but Bulwer seems here to be modeling his own suggestive system: he finds value in Bacon that Bacon himself may very well not have intended. Bacon is useful not just for the limits he places on scientific observation, but also for the imaginative possibilities he engenders, enabling his readers to envision themselves as wiser, better future selves.

The reverence for Bacon that L—— evinces comes after he "grew chilled and dissatisfied with the materialists" and leads him to his own analogical system. He explains his intellectual progress: "Helvetius charmed my fancy—sharpened my intellect—but filled not my soul. Locke, Condillac, alike left me disappointed—and asking solutions to questions which they either dared not answer—or discouragingly evaded." After finding the "Scotch, and . . . the German reasoners" (that is to say, the empiricists and the idealists), likewise inadequate, L—— experiences a sort of enlightenment: "My books were deaf and sealed, but round me was the Universe, and the life of things became my teacher!—Yes—not from metaphysics, but from *analogy* I rebuilt up my crumbling faith,—and became a Philosopher to myself."[38] Those familiar with Gillian Beer's subtle analysis of analogy in *Darwin's Plots* will not be surprised that analogical reasoning was a hallmark of Victorian science, though they will also, perhaps, remember Beer's caution that its status was by no means uncontested. Not only was analogy, as she notes, "part of the armoury of Natural Theology," because it "seemed to provide evidence for a teleological order," but Victorian thinkers, natural theological and materialist alike, understood its epistemological slipperiness: "Its seductively partial applicability, its tendency to suppress all disanalogous elements, means that it can claim more than it proves."[39] So we might read L——'s *Bildung* here merely as Bulwer's Whewellian espousal of natural theology over the "godless" materialism he found so inadequate, but as I will argue in what follows, it is more than this. Analogy is central to the suggestive system that Bulwer develops in *Caxtoniana.*

Caxtoniana: *"Normal" Clairvoyance, Moral Effects, and New Theories*

When Bulwer links Shakespeare's creative genius with Newton's in "On the Normal Clairvoyance of the Imagination," he argues by analogy: the "poetry of philosophers" is like the "philosophy of poets," in that both involve "the gift of seeing through other organs than the eyes," a faculty that is "more or less accurately shared by all in whom imagination is strongly concentred upon any selected object, however distant and apart from the positive experience of material senses."[40] Both creative writers and scientists must be able to imagine what they do not yet see, in the first case in order to describe people, places, and times that are beyond the writer's literal observation and, in the second case, because "men disciplined in the study of severest science, only through reason discover what through imagination they previse."[41] This essay is typical of *Caxtoniana,* and it does its work on several registers at once, linking science, the occult, and literary creativity.

The essay begins by acknowledging the difficulty of proving the existence of real clairvoyance inductively: "So uncertain, indeed, so unreliable, are the higher phenomena ascribed to mesmeric clairvoyance, that the experiments of such phenomena almost invariably fail when subjected to those tests which the incredulous not unreasonably demand."[42] Bulwer hedges his bets somewhat, perhaps unwilling to endorse wholeheartedly occult beliefs that might attract ridicule.[43] But this acknowledgment also provides the setup for his strange argument: if we can't find sufficient evidence to prove mesmeric clairvoyance, then we need look no further than the closest poem or novel to see a species of that same faculty: "Nothing is more frequent among novelists, even third-rate and fourth-rate, than 'to see through other organs than their eyes.' . . . They can describe scenes they have never witnessed more faithfully than the native who has *lived* amid those scenes from his cradle."[44] But it is not merely that the literary imagination is able to supply information that may *seem* credible; rather, the author literally *becomes* clairvoyant, much in the manner of George Eliot's protagonist in *The Lifted Veil* (1859), mystically able to see places he has never been. Bulwer offers himself as proof:

> Just as a chemist who suggests a theory naturally adds to his suggestion the statement of his own experiments, I offer my personal evidence in favour of the doctrine I advance—viz., "that there is nothing so rare as to excite our incredulous wonder in the faculty of seeing 'through other organs than the eyes.'" I have had sometimes to describe minutely, scenes which . . . I had never witnessed. I visited those scenes later. I then examined them, with natural apprehension that I must have committed some notable mistakes to be carefully corrected in any subsequent edition of the work. . . . In no single instance could I ever find, after the most rigid scrutiny, that the clairvoyance of imagination had deceived me.[45]

It is not incidental that Bulwer likens himself to a chemist using his own experiments in support of his theory; as he elsewhere refers to Sir Humphrey Davy's writings, we might imagine that he has in mind Davy's experiments huffing nitrous oxide (with his friend Coleridge) in pursuit of a theory of the properties of gasses. Davy, himself a proponent of Baconianism, certainly blurred the lines between the scientist as objective observer and romantic visionary.[46]

Bulwer's literary clairvoyance, then, does double duty. It lends occult mystique to his "brethren in the masonry of fiction," but at the same time it demystifies mesmeric clairvoyance that, taken by itself, seems "too preternatural, too transcendent for human attainment."[47] Bulwer draws an analogy between "real" clairvoyance and literary creativity, only to merge the two analogues into one phenomenon, in which the literary version is factual proof of the occult. And, to make this move, he draws yet again on analogy: the creative imagination of the writer is like that of the scientist, Newton, who possesses the ability to imagine what he will later discover through experimentation. Bulwer concludes with one last analogy: "The truth really seems to be, that the imagination acquires by custom a certain involuntary unconscious power of observation and comparison, correcting its own mistakes, and arriving at precision of judgment, just as the outward eye is disciplined to compare, adjust, estimate, measure, the objects reflected on the back of its retina."[48] Bulwer transposes the inductive faculties of the scientific "outward eye" to the clairvoyant "gift of seeing through other organs than the eyes," thus inviting his reader to reconceive of both simultaneously.

If analogy enables Bulwer to suggest correspondences among the occult, literary imagination, and inductive science in "On the Normal Clairvoyance of the Imagination," elsewhere in *Caxtoniana* he combines these same elements to theorize the process by which "suggestive" writing and reading generate new ideas. In "On the Moral Effect of Writers," the essay in which he refers to Godwin and also insists upon the value of Newton's alchemical research, Bulwer reflects upon the slippage between the author's intentions and the effect that his or her works may produce. The issue clearly engenders a certain level of anxiety on his own behalf, but it also underscores the similarities between occult knowledge and literary creativity:

> Certain I am that every author who has written a book with earnest forethought and fondly-cherished designs, will bear testimony to the fact, that much which he meant to convey has never been guessed at in any review of his work; and many a delicate beauty of thought, on which he principally valued himself, remains, like the statue of Isis, an image of truth from which no hand lifts the veil.[49]

Tellingly, Bulwer invokes the same imagery here that he uses in *Zanoni* to describe, at different times, the search for occult knowledge and the difficulty of divining an author's meaning in a literary work. In both cases, he refers to Schiller poems ("The Veiled Image at Sais" and "Cassandra") that he himself translated in 1844.[50]

The veiled statue of Isis in *Zanoni*, as in Schiller, represents the perilous allure of occult knowledge. In an epigraph from "The Veiled Image at Sais," Bulwer quotes the reckless "youth" who disregards the priest's warnings and pulls aside the statue's veil (with predictably dire results): "Sey hinter ihm was will! Ich heb ihn auf (Be behind what there may—I raise the veil)."[51] The doomed youth is likened to Glyndon, Zanoni's young protégé who, similarly, disregards his mentor's warnings and grasps at occult knowledge for which he is unprepared. But, Bulwer also deploys the allusion as a less fatal metaphor for the gap between the author's intentions and his work's apprehension by readers. In the "Note" that he appended to all editions of *Zanoni* after 1853, Bulwer tells his readers: "We behold the figure, we cannot lift the veil. The Author himself is not called upon to explain what he designed."[52] In this case, the veiled statue offers more promise than warning, for if an author cannot control the way he is (mis)read, he may take comfort in the idea that future generations of readers may understand his "genius" differently: "Neither Ben Jonson nor even Milton comprehended the sovereign Mastership of Art in Shakespeare. But Shakespeare himself could not have been conscious of his own art. And no writer, whatever his moral object, can foresee what in the course of ages may be the moral effect of his performance."[53] Here author and reader are together "unconscious" of all the potential (and potent) meanings within a text. "*We* behold the figure, *we* cannot lift the veil."

That even Shakespeare was underappreciated in his own time is seemingly a consolatory message for Bulwer, sensitive to criticism and often mocked by his own contemporaries despite his great popularity. But more than that it bespeaks an approach to reading that, as he remarks, "confirm[s] the wisdom of complete toleration to the freedom of all opinion," because we cannot predict what effect any given book may produce on its readers, or what other books and ideas it might generate as a result of its being read: "Had some mistaken benevolence of intention suppressed the publication of Hume's sceptical theories . . . it would have suppressed also all those great arguments for an immaterial soul in man. . . . Kant would have continued in his 'dogmatic slumber;' Reid would have remained in quiet adhesion to Locke; the materialism of Condillac would still be reigning over the schools of France."[54] Materialism begets idealism dialectically. To read (or not read), then, with preconceptions—that is to say, to be "too exclusively anxious about the consistency of system"[55]—is to foreclose the possible benefits that cannot be predicted ahead of time or even necessarily traced after the fact. "We cannot," Bulwer writes, track all the "sources from which we derive the ideas that make us what we are." Indeed, the powerful effects of texts are so powerful as to extend beyond those who actually read them: "Few of my readers may have read Chaucer; fewer still the 'Principia' of Newton. Yet how much poorer the minds of all my readers would be if Chaucer and Newton had never written! . . . But who shall resolve to each individual start the rays of heat and light, whose effects are felt by all, whose nature is defined by none?"[56] Here Chaucer and Newton are to readers as "rays of heat and light," an analogy that suggests (again) that to trace books' effects, one must become like Newton, who undertook just such a study of the nature of light in his *Opticks* (1704). Thus, as I will argue in the coda, Bulwer provides the readers of *Zanoni* with an instructive example of his own text's effects on one reader.

Coda: Zanoni *and Its "Ingenious" Reader*

Bulwer's first occult novel, *Zanoni* tells the tragic story of the eponymous hero, a Chaldean mystic who has mastered the secrets of alchemy. In Naples the immortal Zanoni falls in love, against his better judgment and the advice of his mentor Mejnour, with a beautiful young opera singer, Viola, who cannot comprehend the occult knowledge that could give her the same immortality. Zanoni tries (for everyone's good) to give Viola to his rival for her affections, a callow if promising English gentleman, Glyndon, but Glyndon rejects the offer in order to join Zanoni's and Mejnour's brotherhood, and Zanoni marries Viola. Glyndon fails to adhere to the strict discipline required to follow in Zanoni's footsteps and, encountering the hideous supernatural being, the "Dweller of the Threshold," that guards the knowledge he seeks, Glyndon recoils in horror from Zanoni and Mejnour and his occult pursuits. Glyndon convinces Viola, who has become increasingly terrified of her husband's seemingly magical abilities, to run away with him to Paris, unfortunately on the eve of the Reign of Terror. Events reach their crisis when Zanoni sacrifices his own life to save Viola and Glyndon from the guillotine. Since its publication, *Zanoni* has been interpreted variously as a Godwinian meditation on immortality and human perfectibility, indebted to Godwin's *St. Leon* (1799), a Carlylean critique of the French Revolution, and a romance in the German tradition.[57] It has often been understood, going back to Victorian readers, as expressing Bulwer's "revolt from the chilling materialism of the age,"[58] and, thus, philosophically linked to his second occult novel, *A Strange Story*. *Zanoni* certainly includes evidence aplenty to support all of these readings. But given Bulwer's insistence on the gap between a story's moral and its tendency, we perhaps ought to be cautious in arriving at the "key" to the novel's meaning. Following his and Godwin's injunction to pursue an inductive approach to the "abstruse" science of parsing readers' literary interpretations, my coda is less a reading of *Zanoni* than a reading of the reading of *Zanoni*.

As I mentioned previously, Bulwer's note, appended to the 1853 and subsequent editions of *Zanoni*, pondered the impossibility of readers arriving at one "correct" interpretation of his novel; in language very similar to that which he would use a decade later in his *Caxtoniana* essays, he stipulates that "*Zanoni* is not, as some have supposed, an allegory; but beneath the narrative . . . *typical* meanings are concealed."[59] If it were an allegory, it would be easy to interpret, as "an Allegory is a personation of distinct and definite things." But types are trickier: "A writer who conveys typical meanings, may express them in myriads." And, just as one cannot easily distinguish all the rays of heat and light produced by Chaucer and Newton, so here even the author himself "cannot disentangle all the hues which commingle into the light he seeks to cast on the truth."[60] And, thus, "the author of Zanoni gives . . . no key to the mysteries, be they trivial or important, which may be found in the secret chambers by those who lift the tapestry from the wall."[61] Why, then, a few sentences later, is there just such a key, which must be insufficient to the task of decoding his "typical" novel?

In 1842, a few months after the publication of *Zanoni*, Harriet Martineau offered Bulwer a little something she had written, she claimed, to help her friends who were not as good at interpreting literature as she and who would, therefore, struggle with *Zanoni*. Bulwer was pleased to receive it, and in 1853, he added this key as an appendix to the novel following his note, though without naming Martineau as the author.[62] Robert Wolff reads this decision as Bulwer's anxious desire to insure that he was properly understood by his readers, which then makes his previous disclaimer about the novel's indecipherability insincere. Wolff finds Bulwer's note, then, "disingenuous, defensive, and rather adding to the mystification of *Zanoni* than elucidating it."[63] Disingenuous it may be, but not, I would suggest, as Wolff claims, because Bulwer has Martineau do the exegetical heavy lifting while he declines to endorse or condemn her reading.

As in *Caxtoniana*, in the note Bulwer advocates a "suggestive" mode for reading. Referring to Martineau's key (and speaking of himself in the third person), Bulwer writes: "He leaves it to the reader to agree with, or dissent from, the explanation. 'A hundred men,' says the old Platonist, 'may read the book by the help of the same lamp, yet all may differ on the text; for the lamp only lights the characters—the mind must divine the meaning.' The object of a Parable is not that of a Problem; it does not seek to convince, but to suggest."[64] That Martineau provides a definitive interpretation, confidently reading all the characters as allegorical figures, each clearly representing an Idea (Science, Idealism, Human Instinct, etc.), suggests that she does not understand the distinction that Bulwer is at pains to articulate. An uncharitable interpretation might be that Bulwer includes Martineau's key as an example of misreading—the product of one too much wedded to her system to discern whatever truths may lie in the "minute and near" details.[65] Perhaps this would be to read its inclusion as a mean-spirited in-joke. However, to read the appendix more charitably, we might say that Bulwer uses the opportunity to offer evidence of the gap between moral and tendency. In the novel itself characters presented with the same information, the same books, and the same natural and social phenomena come to widely divergent conclusions: if Zanoni and Mejnour receive profound truths, Glyndon and Viola stubbornly reject the "Universe and the life of things."[66] So, too, *Zanoni*'s actual readers illustrate Bulwer's suggestive system. Bulwer does certainly "cast [his] thoughts freely abroad,"[67] and they do prove valuable insofar as they are generative of Martineau's interpretations, however far afield from his intentions they may or may not be. By including Martineau's key, Bulwer shows what fruits his ideas have produced in one instance, but he also forwards intellectual exchange more broadly, enabling Martineau's ideas to cross-pollinate his own.

To conclude, while we can read Bulwer's suggestive approach to reading as engaging in his culture's widespread interest in scientific methodology in general (and Baconianism in particular), we can also see it as anticipating our own literary-critical theories of unstable texts, interpretive communities, and discursive functions. As Bulwer remarks (clairvoyantly, one might imagine), "Critics, in later times, gain repute by discovering what the author did not mean."[68] Indeed, Bulwer's sense of the afterlives of texts—growing unpredictably, organically in excess of their authors' intentions, generating new connections and ever-changing significations—might remind us of the conclusion to Foucault's seminal essay "What Is an Author?," in which he suggests that the questions to ask about a text are not those that refer to the author-as-origin, the questions that reassert "the privileges of the subject," but instead those that analyze the subject "as a variable and complex function of discourse."[69] Given the many variant translations of Foucault's essay, which provide differing glosses of his theory of the "author-function," Bulwer's comment on future interpreters' license holds as true for Foucault's exegetes as it did for his own, or Bacon's for that matter.[70] In Bulwer's suggestive system we can find, not just a strange instance of the intersection of Victorian scientific, literary, and occult discourses but a nascent theory to describe the "proliferation of meaning" that arises in such intersections.[71] Of course, to compare Bulwer's eclectic theory-building in the all-but-forgotten *Caxtoniana* and *Zanoni* with a pivotal essay by one of poststructuralism's foremost theorists may strain credulity (even if Foucault does mention Bacon and Newton in "What Is an Author?"); nonetheless, Bulwer's eclectic repurposing of Baconian induction should serve as a reminder that undertaking the intellectual labor required to envision unlikely connections is worthwhile precisely because "Thought is valuable in proportion as it is generative."[72]

Notes

1. Edward Bulwer-Lytton, "On Essay-Writing," in *Caxtoniana: A Series of Essays* (1863; London: George Routledge and Sons, 1875), 150.

2. Ibid.

3. Ibid.

4. Bulwer studied with Whewell at Trinity College, Cambridge, during which time he discussed Bacon and induction frequently with the then tutor and fellow. According to T. H. S. Escott, "Whewell was then meditating his various books on the inductive sciences and on moral philosophy in England. These formed the subject of many conversations with Bulwer." *Edward Bulwer, First Baron Lytton of Knebworth: A Social, Personal, and Political Monograph* (London: George Routledge and Sons, 1910), 45.

5. Bulwer helpfully provides the provenance for his allusion to Bacon in a footnote: "Bacon's Essay on Atheism. This quotation is made with admirable felicity and force by Dr. Whewell, page 378 of Bridgewater Treatise, On Astronomy and General Physics considered with Reference to Natural Theology." *A Strange Story* (London: Sampson and Low, 1862), 2:67.

6. Ibid., 64.

7. William Godwin, *The Enquirer* (London: G. G. & J. Robinson, 1797), 136.

8. Ibid., 133.

9. Ibid., vi.

10. For an excellent broad overview of Bacon and his influence, see Paolo Rossi, "Baconianism," in *Dictionary of the History of Ideas: Studies of Selected Pivotal Ideas*, ed. Philip P. Wiender, 5 vols. (New York: Scribner's, 1973), 1:172–79.

11. Godwin, *Enquirer*, vii, v.

12. Edward Bulwer-Lytton, "On the Moral Effect of Writers," in *Caxtoniana*, 112.

13. Ibid., 123–24.

14. Edward Bulwer-Lytton, "Hints on Mental Culture," in *Caxtoniana*, 101.

15. Lawrence Poston, "Bulwer's Godwinian Myth," in *The Subverting Vision of Bulwer Lytton: Bicentenary Reflections*, ed. Allan Conrad Christensen (Newark: University of Delaware Press, 2004), 78–79. See also Andrew Brown, "Metaphysics and Melodrama: Bulwer's *Rienzi*," *Nineteenth-Century Fiction* 36, no. 3 (1981): 266.

16. Joseph I. Fradin, "'The Absorbing Tyranny of Every-Day Life': Bulwer-Lytton's *A Strange Story*," *Nineteenth-Century Fiction* 16, no. 1 (1961): 2.

17. Christopher Lane, "Bulwer's Misanthropes and the Limits of Victorian Sympathy," *Victorian Studies* 44, no. 4 (2002): 598.

18. Margaret F. King and Elliot Engel remark: "Since he is hardly a genius of the first order, Bulwer's humorless pretensions, as reflected in his turgid, pompous prose, do encourage a certain pleasure in denigrating him." "The Emerging Carlylean Hero in Bulwer's Novels of the 1830s," *Nineteenth-Century Fiction* 36, no. 3 (1981): 277.

19. Marie Mulvey-Roberts, "Fame, Notoriety and Madness: Edward Bulwer-Lytton Paying the Price of Greatness," *Critical Survey* 13, no. 2 (2001): 115. The scholarly disclaimers about the relative merits of Bulwer's works that preface so much scholarship would make a study in themselves. I agree with Peter W. Sinnema's assessment that "Lytton's widely accepted status as a third-rate Victorian novelist justifies the confidence with which critics declaim upon his monotonous interest in metaphysical perplexities and his reduction of characters to predictable types without, generally, any attempt to demonstrate how such 'doctrine[s]' . . . are manifested in particular prose narratives." "Domesticating Bulwer-Lytton's Colonial Fiction: Mentorship and Masculinity in the Caxtons Trilogy," *English Studies in Canada* 26, no. 2 (2000): 155. There are, of course, exceptions to the scholarly neglect of (or, worse, contemptuous attention to) Bulwer and his works. John Coates reads Bulwer, in *Zanoni* in particular, as formative rather than reflective of "the ideas of his age." "*Zanoni* by Bulwer-Lytton: A Discussion of Its Philosophy and Its Possible Influences," *Durham University Journal* 76, no. 2 (1984): 224. For the most thorough treatment of Bulwer's occult fiction see Robert Lee Wolff, *Strange Stories and Other Explorations in Victorian Fiction* (Boston: Gambit Press, 1971). For excellent and varied treatments of many of Bulwer's works, see Allan Conrad Christensen, ed., *The Subverting Vision of Bulwer Lytton: Bicentenary Reflections* (Newark: University of Delaware Press, 2004). See also Felicia Bonaparte, "The Mythification of History and the Historicizing of Myth: Bulwer, Wagner, and *Rienzi*," *Clio* 37, no. 3 (2008): 365–93; Gavin Budge, "Mesmerism and Medicine in Bulwer-Lytton's Novels of the Occult," in *Victorian Literary Mesmerism*, ed. Martin Willis and Catherine Wynne (Amsterdam: Rodopi Press, 2006), 39–59; Nicholas Daly, "The Volcanic Disaster Narrative: From Pleasure Garden to Canvas, Page, and Stage," *Victorian Studies* 53, no. 2 (2011): 255–85; Mark Knight, "'The Haunted and the Haunters': Bulwer Lytton's Philosophical Ghost Story," *Nineteenth-Century Contexts* 28, no. 3 (2006): 245–55; Ted Underwood, "Historical Difference as Immortality in the Mid-Nineteenth-Century Novel," *Modern Language Quarterly* 63, no. 4 (2002): 441–69; Stanwood Walker, "Redeeming the 'City of the Dead': Metaphysical Fiction, Touristic Fantasy, and the Historical Other in Bulwer's *The Last Days of Pompeii*," *Victorian Newsletter* 116 (Fall 2009): 27–51.

20. Bulwer-Lytton, "On Essay-Writing," 150.

21. Edward Bulwer-Lytton, "On the Normal Clairvoyance of the Imagination," in *Caxtoniana*, 40.

22. Bulwer-Lytton, "On Essay Writing," 150.

23. Knight, "'Haunted and Haunters,'" 245.

24. George Levine, *Dying to Know: Scientific Epistemology and Narrative in Victorian England* (Chicago: University of Chicago Press, 2002), 20. Although neither discusses Baconianism in any detail, for excellent accounts of Victorian scientific methodologies see also George Levine, *Realism, Ethics, and Secularism: Essays on Victorian Literature and Science* (Cambridge: Cambridge University Press, 2008); Christopher Herbert, *Victorian Relativity: Radical Thought and Scientific Discovery* (Chicago: University of Chicago Press, 2001).

25. Jonathan Smith, *Fact and Feeling: Baconian Science and the Nineteenth-Century Literary Imagination* (Madison: University of Wisconsin Press, 1994), 13.

26. Richard Yeo, "An Idol of the Market-Place: Baconianism in Nineteenth-Century Britain," *History of Science* 23, no. 3 (1985): 252.

27. Smith, *Fact and Feeling*, 13.

28. Wolff, *Strange Stories*, 90.

29. Henry George Atkinson and Harriet Martineau, *Letters on the Laws of Man's Nature and Development* (London: John Chapman, 1851), 139.

30. The review appeared in two parts. See [G. H. Lewes], review of *Letters on the Laws of Man's Nature and Development*, by Henry George Atkinson and Harriet Martineau, *Leader*, March 1, 1851, 201–3; [G. H. Lewes], "Second Notice," review of *Letters on the Laws of Man's Nature and Development*, by Henry George Atkinson and Harriet Martineau, *Leader*, March 8, 1851, 227–28. He remarks acerbically in the latter of the authors' proffered evidences of mesmerism and clairvoyance: "They must excite ridicule such as will throw discredit upon the work; and they cannot be accepted by any cautious mind, however predisposed to believe in the marvels of mesmerism. And, in saying this, we are not prejudging mesmerism" (227).

31. And Martineau is certainly better remembered for her more prosaic works like *Illustrations of Political Economy* (1832–34) and *The Positive Philosophy of Auguste Comte* (1853) than for her collaboration with Atkinson.

32. Edward Bulwer-Lytton, "On the Spirit in Which New Theories Should Be Received," in *Caxtoniana*, 136.

33. Ibid., 138.

34. Ibid.

35. Smith, *Fact and Feeling*, 16. For discussions of the ways Baconianism was associated with political and philosophical controversy see also Antonio Pérez-Ramos, *Francis Bacon's Idea of Science and the Maker's Knowledge Tradition* (Oxford: Clarendon Press, 1988); Yeo, "Idol of the Market-Place," 251–98.

36. G. H. Lewes, *Comte's Philosophy of the Sciences: Being an Exposition of the Principles of the "Cours de Philosophie Positive" of August Comte* (London: George Bell and Sons, 1897), 104–5.

37. Edward Bulwer-Lytton, "The New Phaedo, or, Conversations on Things Human and Divine with One Condemned," in *The Student: A Series of Papers*, vol. 2 (London: Saunders and Otley, 1836), 159. L—— often appears to be a projection of Bulwer, in places narrating fictionalized biographical details that were Bulwer's own. "The New Phaedo" first appeared in installments in the *New Monthly Magazine* beginning in 1830 and was collected, along with other essays from the same journal, in volume form in 1835.

38. Ibid., 240–41. For a comprehensive account of the Victorian empiricism/idealism debates see Peter Garratt, *Victorian Empiricism: Self, Knowledge, and Reality in Ruskin, Bain, Lewes, Spencer, and George Eliot* (Madison, NJ: Fairleigh Dickinson University Press, 2010).

39. Gillian Beer, *Darwin's Plots: Evolutionary Narrative in Darwin, George Eliot and Nineteenth-Century Fiction*, 2nd ed. (Cambridge: Cambridge University Press, 2003), 76–77. See also Michael Carignan, "Analogical Reasoning in Victorian Historical Epistemology," *Journal of the History of Ideas* 64, no. 3 (2003): 445–64; Jonathan Farina, "Dickens's As If: Analogy and Victorian Virtual Reality," *Victorian Studies* 53, no. 3 (2011): 427–36; Kevin Lambert, "The Uses of Analogy: James Clerk Maxwell's 'On Faraday's Lines of Force' and Early Victorian Analogical Argument," *British Journal*

for the History of Science 44, no. 1 (2011): 61–88. Kaja Silverman offers a much more utopian, if less historically grounded, account of analogy as an epistemological (and ethical) alternative to individualism in which individuals are "linked to each other through reversible and ontologically equalizing similarities." *Flesh of My Flesh* (Stanford, CA: Stanford University Press, 2009), 4.

40. Bulwer-Lytton, "On the Normal," 38–39.

41. Ibid., 40.

42. Ibid., 35.

43. Leslie Mitchell's biography paints him as acutely sensitive to ridicule, particularly (though by no means exclusively) in relation to his occult interests: "He investigated phenomena at the risk of his reputation, calling on parallel, scientific experiments to justify his actions. He was very aware that, in this respect, he was walking a tightrope and that there was every risk of falling into complete nonsense." *Bulwer Lytton: The Rise and Fall of a Victorian Man of Letters* (London: Hambledon and London, 2003), 144.

44. Bulwer-Lytton, "On the Normal," 39.

45. Ibid.

46. Smith reads Davy in relation to the Romantic poets, arguing, "Wordsworth gives poetry a scientific basis in fact, in careful observation and exact description, which is then imaginatively transformed in an inductive manner. . . . Davy contends that science utilizes this same poetic transformation of empirical data into the elegant laws of nature which in turn reinforce the beauty and harmony of the natural world." *Fact and Feeling*, 58.

47. Bulwer-Lytton, "On the Normal," 39, 36.

48. Ibid., 41.

49. Bulwer-Lytton, "On the Moral," 118.

50. See Friedrich Schiller, "The Veiled Image at Sais," and "Cassandra," in *Schiller's Poems and Ballads*, trans. Edward, Lord Lytton (1844; New York: George Routledge and Sons, 1887), 61–63, 98–102.

51. Edward Bulwer-Lytton, *Zanoni* (1842; London: G. Routledge, 1856), 140.

52. Ibid., 301.

53. Bulwer-Lytton, "On the Moral," 114.

54. Ibid., 124.

55. Godwin, *Enquirer*, vi.

56. Bulwer-Lytton, "On the Moral," 125.

57. For discussions of Bulwer's various antecedents, see especially Bonaparte, "Mythification of History," 365–93; Coates, "*Zanoni*," 223–33; King and Engel, "The Emerging Carlylean," 277–95; Lane, "Bulwer's Misanthropes," 597–624; and Lawrence Poston, "Beyond the Occult: The Godwinian Nexus of Bulwer's *Zanoni*," *Studies in Romanticism* 37, no. 2 (1998): 131–61.

58. William Mathews, introduction to *Zanoni*, in *Novels and Romances of Edward Bulwer Lytton*, vol. 23, ed. William Mathews (Boston: Little, Brown, 1893), xi. See also Budge, "Mesmerism and Medicine," 39–59.

59. Bulwer-Lytton, *Zanoni*, 301.

60. Ibid.

61. Ibid., 302.

62. Wolff, *Strange Stories*, 208ff.

63. Ibid., 217.

64. Bulwer-Lytton, *Zanoni*, 302.

65. Godwin, *Enquirer*, vi.

66. Bulwer-Lytton, "New Phaedo," 241.

67. Bulwer-Lytton, "On Essay-Writing," 150.

68. Bulwer-Lytton, "On the Moral," 117.

69. Michel Foucault, "What Is an Author?," in *The Foucault Reader*, ed. Paul Rabinow (New York: Pantheon, 1984), 117–18.

70. For a list of variant translations of "What Is an Author?" see Richard Lynch, "Michel Foucault's Shorter Works in English," Michel-foucault.com, last modified October 30, 2010, http://www.michel-foucault.com/bibmf/. For a discussion of these variants, see Joseph Long, "A Company of Shades: Subject and Authorship in Samuel Beckett's Prose," in *Subject Matters: Subject and Self in French Literature from Descartes to the Present*, ed. Paul Gifford and Johnnie Gratton (Amsterdam: Rodopi, 2000), 115–16; see also Richard Lynch, "Two Bibliographical Resources for Foucault's Work in English," *Foucault Studies* 1 (December 2004): 71–76.

71. Foucault, "What Is an Author?" 119.

72. Bulwer-Lytton, "On Essay Writing," 150.

CHAPTER 12

Psychical Research and the Fantastic Science of Spirits

L. Anne Delgado

In her 1918 review of American critic Dorothy Scarborough's *The Supernatural in Modern Literature*, Virginia Woolf observed that "the great increase of the psychical ghost story in late years to which Miss Scarborough bears witness, testifies to the fact that our sense of our own ghostliness has much quickened."[1] Woolf was commenting upon a literary trend that had taken shape during the late nineteenth and early twentieth centuries and was inspired by the exhaustive investigations of the unseen world conducted by the British Society for Psychical Research (SPR), an organization established in 1882, nearly forty years prior to her review. Woolf's observation is interesting in that it illustrates the degree to which the ghost and the Victorian ghost story had been reformed, and perhaps rehabilitated, by the field of psychical research. Indeed, the psychical ghost story would not have manifested at all had it not been for the SPR's concentrated attempt to account for the new ghost's odd habits, and this new ghost was a curious entity. In contrast to the uncanny tendencies of the old ghost—with its tangle of chains and tenebrous nature—the SPR's ghost cut a more substantial figure. Assembled from a fat catalog of ghostly encounters that the Society solicited from the Victorian public, the SPR's ghost was ultimately a manifestation shaped by statistical data. For the late Victorians, and through the efforts of the SPR, this ghost became real in a way that seemed unprecedented, and its intrusions were meticulously documented. Like Woolf, who describes a quickened sense of ghostliness at the dawn of the new century, the Victorians' sense of ghostliness had been revived by both the spiritualist movement and the development of psychical research. In contrast to Woolf's, and perhaps Henry James's, specters—entities that reflected ambiguous iterations of the self—the psychical and spiritualist ghost was expected to perform in such a way that its performances could be documented. While the spiritualists relied upon these performances as a means of sustaining faith, psychical researchers interpreted reports of spectral patterns of performance as evidence that might sustain their new science of spirits. Their task was a daunting one. Not only had spiritualism reshaped the ways in which the ghost performed, but an extensive literary history also dictated the ways in which this ghost might behave. In their attempts to wrest the idea of the ghost from the realm of faith and fantasy, the SPR sought to transform it into an entity that could be measured, analyzed, and cataloged. However, as this chapter illustrates, the SPR's ghost radically revealed the degree to which the ghost, and the nature of ghostly encounters, was bound by literary precedents, particularly as it emerged in the 1886 *Phantasms of the Living*, a groundbreaking investigation of telepathy and spectral phenomena.[2]

In this chapter, I use the term "psychical ghost" to refer to spectral entities as they were defined by the SPR in their attempts to reorient Victorian understandings of the phantasmal. On the one hand, the term captures the antithetical aims of such research: it illustrates the tension between the psychical, or natural, world and the ghostly, or supernatural, world that dogged the Society in its attempts to articulate a spectral theory. On the other hand, the term is situated within a continuum of ideas that rarely strays from its fantastic origins. Since the Society's study had been built upon anecdotal accounts of ghostly encounters drawn from the Victorian public, the nature of these "true stories" often mirrored ghostly encounters described in popular fiction. In this sense, the psychical ghost embodied the paradoxical and distinctly Victorian reverence for empirical data and spectral fantasy. The term further accounts for the wholly fantastic conception of the ghost that—as Virginia Woolf's comment illustrates—continued to haunt the psychical researcher's new science of spirits.

Several critics have examined the ideas that informed and identified the transformation of the ghost, particularly as it emerged in the nineteenth century. In this chapter, I hope to push this particular field of inquiry into new territory. In his *Ghost-Seers, Detectives, and Spiritualists: Theories of Vision in Victorian Literature and Science* (2010), Srdjan Smajic examines the ways in which fictional representations of the nineteenth-century ghost revealed the unsteady nature of both human perception and the knowledge that perception itself produces.[3] These limitations are certainly evident in SPR's *Phantasms of the Living*. However, as this chapter suggests, the ways in which the SPR utilized witness accounts of the ghost in establishing a new body of knowledge reveal the transitional nature of the psychical ghost, particularly as it emerged in *Phantasms of the Living*. In *The Invention of Telepathy* (2002) Roger Luckhurst refers to this text as one "whose title announced that the telepathic theory of hallucinated figures from agent to percipient discarded the spirits of the dead on which Spiritualism relied."[4] In contrast to the SPR's repudiation of the ghost and Luckhurst's characterization of psychical research as a field of study that seemed to escape the mire of spiritualism through the "invention of telepathy," I argue that such research actually reflected popular conceptions of the unseen world. While the idea of telepathy structured the Society's theories, the cases themselves are not necessarily relegated to telepathy. Furthermore, examination of the SPR's ghost qua veridical hallucination also places it within a context that is not, as Shane McCorristine argues in his excellent *Spectres of the Self: Thinking about Ghosts and Ghost-Seeing in England, 1750–1920*, merely a "modern conception of the ghost as reflective of the haunted nature of the self."[5] Instead, the psychical ghost marks an important transitional stage in Victorian understandings of the ghost. Thus, I argue that the larval specter that we encounter in *Phantasms of the Living* is in the midst of transforming from a literary entity into a scientific one. The subsequent development of this entity, particularly after the critical scorn and public curiosity that the study attracted, led to its fragmentation: on the one hand it was reabsorbed by the literary tradition, and on the other hand it became a metaphor for the modern self.

The Science of Spirit and Spectral Taxonomies

The Society, established in 1882 by the philosopher and Cambridge scholar Henry Sidgwick, was organized around the scientific study of supernatural phenomena and developed in tandem with heterodox spiritualities of the period like spiritualism and theosophy.[6] While the emergence of the SPR may seem somewhat unrelated to the surge in heterodox spiritual activity in the last decades of the century, the Society was inspired by a similar dissatisfaction with the limitations of orthodox theologies. In his biography of Henry Sidgwick, Bart Schultz writes that "it was the battering dealt his Anglican beliefs during his years of 'storm and stress,' when he came to struggle so with the entire issue of the evidence for miraculous happenings, that pushed him to accord a truly cosmic significance to [the paranormal]."[7] On the opposite side of religious orthodoxy sat the specter of materialism. F. W. H. Myers, a former student of Sidgwick's and founding member of the SPR, poetically characterized materialism as "a dull pain borne with joyless doggedness, [that] sometimes flashed into a horror of reality that made the world spin before one's eyes,—a shock of nightmare-panic amid the glaring dreariness of the day."[8] In an effort to mitigate the horrors of materialism, Sidgwick's group, composed primarily of upper-middle-class gentlemen, devoted its time and labor to developing a science of the spirit. Members of the Society included eminent Victorians with diverse political sympathies and interests. The future prime minister and conservative politician Arthur Balfour was a member. Other members included the British poet Alfred Tennyson; Charles Dodgson, otherwise known as Lewis Carroll; the physicists Oliver Lodge and William Barrett; and the chemist William Crookes. These researchers developed a field of inquiry that had far-reaching impact upon late Victorian and Edwardian culture and letters.[9]

In 1883, one year after the SPR was formed, the organization's Literary Committee—a group tasked "with the collection and collation of . . . materials"[10]—began work on *Phantasms of the Living*, a volume that was to become a compendium of what Myers termed "supernormal" events, or phenomena "which are beyond what usually happens" and exhibit "the action of laws higher, in a psychical aspect, than are discerned in action in everyday life."[11] In addition to enumerating various cases of supernormal phenomena, the researchers brought the ghost up to date, recasting it as a "veridical hallucination," a "truth-telling" manifestation that "correspond[ed] to some action which [was] going on in some other place or on some other plane of being," as opposed to "meaningless fictions of an over-stimulated eye or brain."[12] The labor that went into producing this fourteen-hundred-page investigation was unevenly divided among Myers, Frank Podmore, one of the founders of the socialist Fabian Society as well as one of the SPR's early skeptics, and Edmund Gurney, the Society's honorary secretary—as man whose charm and intellectual spirit was said to have inspired George Eliot's fictional hero Daniel Deronda.[13] Gurney's income allowed him to channel all of his energies into psychical research, and he conducted interviews, accumulated data, and compiled the case histories that laid the foundation of this investigation. Gurney threw himself into his labors, often working ceaselessly amid a sea of raw data that included firsthand accounts of thought transference, spontaneous telepathy, and what he called "borderland" cases in which the witness, or percipient, experienced telepathic communication in the moments between sleeping and waking. The bulk of these accounts arrived in the form of letters from a public eager to share personal stories of paranormal experience and in response to requests Gurney published in papers like the *Standard*, the *Pall Mall Gazette*, the *Daily News*, and the *Liverpool Mercury*.[14] As far as the researchers were concerned, the collection and collation of the material was deliberate, methodical, and conducted in the "'dry light' of a dispassionate search for truth."[15] In their pursuit of the truth, the investigators introduced what appeared to be a methodological approach that was guided by a scientific spirit. In testing the reality of thought transference, for example, researchers placed a screen between two subjects—one of them hypnotized—and observed the reactions of the hypnotized subject to stimuli experienced by his nonentranced partner. As for their attempts to verify the existence of other types of psychic phenomena, the researchers collected, corroborated, and interpreted volumes of news stories as well as historical and personal accounts concerning, among other things, deathbed hallucinations and prophetic dreams. The researchers adopted these ostensibly empirical methods of research in order to legitimize their theories of the unseen world. What made the Society's approach to such phenomena scientific was their attempt to organize these phenomena into taxonomic categories. Phantasms were identified by their modes of transmission, which included "visual" and "auditory" cases, sane hallucinations, and telepathic dreams.

In an effort to establish a science of spirits, and in a telling attempt to reorient the ways in which Victorians saw the ghost, the SPR preemptively sought to distinguish these apparitions from fictional ones. The Committee on Haunted Houses, which included Frank Podmore and Hensleigh Wedgwood, Charles Darwin's cousin, submitted the following statement in 1884 in their second report:

> In the magazine ghost stories, which appear in such numbers every Christmas, the ghost is a fearsome being, dressed in a sweeping sheet or shroud, carrying a lighted candle, and squeaking dreadful words

> from fleshless lips. It enters at the stroke of midnight, through the sliding panel, just by the blood stain on the floor, which no effort ever could remove. Or it may be only a clinking of chains, a tread as of armed men, heard whilst the candles burn blue, and the dogs howl. These are the ghosts of fiction, and we do not deny that now and then we receive, apparently on good authority, accounts of apparitions which are stated to exhibit some features of a sensational type. Such cases, however are very rare, and must for the present be dismissed as exceptional.[16]

The committee further separated the ghosts of fiction from the psychical ghost by emphasizing the "casual and objectless nature" of the latter.[17] Ten years later, the folklorist, novelist, and future president of the SPR Andrew Lang further confirmed the committee's characterization of the psychical phantom, noting that it "never speaks. It has no message to convey, or, if it has a message, it does not convey it."[18] For Lang and the above-noted researchers, the psychical ghost was an entity that lacked narrative purpose and, in this respect, had very little in common with the fictional ghost.[19] Indeed, the phantasms that Gurney and his co-researchers endeavored to define were not ghosts at all but telepathic projections of the living. Even the phantasms of the dead, or what the researchers referred to as "crisis apparitions," were manifestations of telepathic rapport between the witness and the decedent. Furthermore, the researchers considered such manifestations valid only if they occurred within twelve hours of the decedent's passing. This window of performance ostensibly allowed the researchers to separate such "phantasms of the living" from fictional and spiritualist "phantasms of the dead."

However, in spite of the SPR's attempts to redefine spectral phenomena as evidence of telepathic rapport, their phantasms were neither entirely purposeless nor bound by the new rules of psychical research. Instead, stories of these encounters, in many ways, reflected the desires of their witnesses and bore a striking resemblance to both the "true ghost story" and the fictional ghost, the former a species found in the works of mid-Victorian writers like Catherine Crowe, about whom I'll say more later, and the latter a type defined by the much more elastic parameters of fiction. In gathering evidence for the existence of veridical hallucinations, Gurney, Myers, and Podmore relied upon specific, recurrent narratives to establish the truth about the supposedly real, but problematically invisible, world. The accounts accumulated by these researchers, perhaps as a testament to their veracity, included minute and irrelevant details that mirrored unsensational Victorian lives. Nevertheless, while the narratives compiled and published by the psychical researchers were not purported to be creative endeavors, they were curiously similar in content to fictional ghostly encounters. In one story from a letter submitted to the committee on June 29, 1884, a woman describes her childhood encounter with what she believes to be the ghost of her brother, "an officer in the 16th Lancers, then quartered in Madras." According to her account, the ghost leaned over to kiss her, then made "signs to me not to speak."[20] Initially overjoyed by her brother's surprise return from India, she becomes disappointed when she learns that he hasn't returned home. The family is later informed of his death from "jungle fever," and the percipient, calculating the day of her brother's death and his spectral appearance, writes, "[I] put two and two together . . . and found that, as I then and now firmly believe, my favorite brother came to me at the hour of his death."[21] In corroborating her own story—she checks her "childish diary" with the military record of her brother's death—the woman establishes "evidence" that supports her account of her brother's spectral return.

In another story, a woman, who, in view of her "position of responsibility," asked to remain anonymous, described seeing what appeared to be the ghost of her absent husband scurrying to her bedside and then disappearing. The woman notes that the room was lit and that, although she was in bed, she had not yet fallen asleep. The woman's husband had been in Australia for some time and had "been an invalid for years."[22] Despite his illness, however, the woman tells the researchers that she had received no word of his condition worsening. A short time after seeing this apparition, what the percipient came to regard as a vision, she hears of his death and, like the woman in the preceding story, discovers that the time of the vision corresponds to the approximate time of her husband's death. The woman's story follows a pattern of revelation and corroboration, narrative patterns that resemble those found in the fictional ghost stories of writers like Mrs. Henry Wood and Rhoda Broughton. The authenticity of such encounters depends upon the almost incidental manifestation of the ghost. The spectral experience, however, is validated through the inclusion of temporal information—times and dates of the encounter coupled with times and dates of the decedent's passing—and corresponding accounts. It was this corroborative data that formed the foundation of the SPR's evidence, a foundation that would eventually undermine the scientific viability of the psychical ghost. For example, in one of the cases Gurney included, a Reverend J. A. Macdonald describes the apparitional encounter of one of his parishioners:

> During the last illness of Mr. William Jackson, of Otley . . . the little son of his daughter sickened and died. Wishing not unnecessarily to disquiet the good man, this sad event was withheld from him. He was full of holy joy, and recognised the presence in his chamber of a number of his relatives who had departed this life in the triumph of faith. He pointed them out in succession—this is so-and-so, and there such another. In the course of this proceeding he suddenly started with surprise, for he discovered his grandson also among the heavenly company.[23]

This faith-affirming story may have been enough for the reverend. However, it was not enough for Gurney, who asked MacDonald for material evidence that might confirm the authenticity of the encounter. MacDonald wrote back and told Gurney that although Mr. Jackson's daughter sent him a written account of this vision, he "destroyed her letter, never dreaming of a Society for Psychical Research."[24] Throughout *Phantasms,* the researchers referred to letters, such as MacDonald's, that might have confirmed apparitional or telepathic phenomena had they been saved. This dearth of evidence became a focal point for many of the book's critics. The summer after *Phantasms of the Living* appeared, lawyer and church historian A. Taylor Innes responded with a "cross-examination of certain phantoms"[25] in the pages of the *Nineteenth Century.* Innes argued that the problem with the study was in its lack of material evidence. Gurney later rebuffed Innes's critique, claiming that the stories themselves were ultimately more significant than the missing data. However, without material corroboration, the narratives failed to confirm anything other than their own existence as stories. As a compilation of "true" ghost stories, however, the SPR's *Phantasms of the Living* would eventually reveal its true potential. Indeed, the chilling element of what came to be known as the psychical ghost story was sharpened by the fact that it had been verified by psychical research.

Of Psychical Ghosts and Psychical Ghost Stories

The ghosts that appear in Gurney's collection, perhaps as a measure of consolation to their witnesses, ultimately return to their homes. To a degree, the psychical ghost restores a sense of domestic and national order: husbands return to wives, sons return to mothers, and soldiers return to their homeland. Ultimately, these apparitions of the deceased, along with their physical bodies, are not abandoned, nor do they remain, in spirit, abroad. Although the veridical hallucination refuses to explain itself, the narratives that such visitations inspire nevertheless reveal an inclination on the part of their authors to entertain fantasies that restore some logic to life. In this respect, the psychical ghost was not so different from the fictional ghost. Contrary to what the SPR's Committee on Haunted Houses had earlier claimed, the psychical ghost had a purpose.

These accounts were also influenced by folkloric conceptions of spectral encounters and bore some similarity to the ghosts found in earlier Victorian collections like the enormously popular *The Night-Side of Nature* (1848), British writer Catherine Crowe's popular study of ghosts, and the American spiritualist Robert Dale Owen's *Footfalls on the Boundary of Another World* (1860).[26] Crowe's work, a philosophical exploration of apparitions, included tales of materializing spirits who offered reassuring evidence of a lively spirit world. But the book also included seemingly purposeless ghosts that simply appeared to bewildered witnesses and disappeared seconds later. Like the "psychical" ghost, Crowe's specters manifested as "crisis apparitions." Shortly after the appearance of the ghost, witnesses soon heard news of their loved one's death. With the appearance of *Phantasms of the Living*, the fictional ghost had been transformed into a psychical one, and Gurney's "census" of veridical hallucinations converted ghostly encounters into evidential narratives. Nevertheless, such supernormal occurrences, although compiled in the detached manner of a sociological sampling, often read like Crowe's and Owen's "true" ghost stories, a genre unsurprisingly reanimated, after the publication of *Phantasms*, in later works like W. T. Stead's *Real Ghost Stories* (1891), Lang's *Book of Dreams and Ghosts* (1897), and the Irish author Elliott O'Donnell's *Byways of Ghostland* (1911).[27] The tales themselves, to say nothing of their popular and largely unrefined sources, however, seemed potentially damning as scientific artifacts. The grim nature of the subject of *Phantasms of the Living* led Oliver Lodge, coinventor of the wireless telegraph and Gurney's colleague in psychical research, to comment: "The book struck me as a meaningless collection of ghost-stories which he was classifying and arranging. . . . Attention to such gruesome tales seemed to me a futile occupation for a cultivated man."[28] In Lodge's opinion, Gurney's absorption in such ghost stories threatened to destroy the credibility of both the researcher and the study.[29]

Unsurprisingly, the SPR's attempts to transform the ghost into a subject of scientific study were at times met with some resistance if not outright hostility from Victorian journalists and critics. The aforementioned Andrew Lang, whose sympathy for the research conducted by the SPR was often divided, argued that the ghost, once endowed with the pleasurable capacity to inspire fear, had now "degenerat[ed] into [a] scientific exercise."[30] Shortly after the publication of *Phantasms of the Living*, a reviewer from the *Daily News* commented upon the length of the text and the labor with which the book taxed its reader: "Even in the present age of steam and electricity the human mind can scarcely master these twelve hundred pages in less than a year or thereabouts. [The authors] stave off the mere lover of ghost stories (with whom we confess our sympathy) by a long preliminary dissertation on telepathy of all kinds."[31] Gurney, perhaps anticipating the disappointment of readers who expected a book full of ghost stories, describes the narratives themselves as "very unexciting—monotonous amid all their variety—as different from the *Mysteries of Udolpho* as from the dignified reports of a learned Society, and far more likely to provoke slumber . . . than to banish it."[32]

The SPR's ghost, at least for Lang and the reviewer from the *Daily News*, had been shorn of the common pleasures it had once afforded readers. In this sense, it would appear that the SPR's methodical treatment of the ghost limited its fictional possibilities. Other reviewers of the work derided the construction of the psychical ghost as one that was built upon "the vaporings of hysterical monomaniacs or the cunningly devised fables of rogues."[33] The SPR itself had been regarded with some suspicion. The *Saturday Review* often referred to the new organization as the "spookical Society,"[34] while a writer from the *Pall Mall Gazette* cautioned its readers that "belief in ghosts, in witchcraft, in second-sight, and all the rest of it is a continuous inheritance of our race from a very remote and savage period," adding that the enterprise of psychical research might invite "dangerous trains of thought."[35] The *Pall Mall Gazette*'s characterization of psychical research itself seems haunted by the possibility that such discussions might trigger dormant, atavistic tendencies among members of the reading public.

But it was the *story* of ghosts in *Phantasms*, rather than the telepathic *theory* of ghosts, that the public most ardently desired. The appearance of the new psychical ghost was timely one, coinciding with the seasonal run of Christmas annuals and issues—what W. T. Stead termed the "ghost season."[36] The *Daily News*, despite claiming that *Phantasms of the Living* only featured "very dull ghosts," nevertheless reproduced some of the ghost stories for its readers.[37] Similarly, in an otherwise acerbic article that appeared on the heels of the book's publication, George Bernard Shaw wrote, "There is no affecting to ignore the public cry of 'Never mind whether the stories are true: let's hear some of them,'" and afterward reproduces four stories that are "most satisfactorily corroborated."[38] Shaw further notes that eliciting "true" ghost stories from a public infatuated with fictional phantoms would inevitably attract witnesses who were "obscurely epileptic or hysterical persons, incorrigibly conceited and mendacious" and "strongly addicted to the miraculous." Unlike the critics or the psychical researchers, the public had no need for corroboration or truth; it simply demanded the stories themselves.[39] Other publications were similarly inclined to pour this new psychical spirit into old bottles. Although the author of "Phantasms of the Living: A New Apology for Ghosts" initially derides the researchers and their methods, he nevertheless includes a sample of Gurney's stories. Since the stories themselves were what readers really desired, the *Derby Mercury* reproduced cases from the book while barely introducing or critiquing the SPR's material or its methods of research.[40] In spite of its numerous critics, psychical investigation and, more specifically, the psychical ghost, had aroused the public's interest.

In his review of *Phantasms of the Living*, William James, largely confident that the phenomena Myers, Podmore, and Gurney compiled were legitimate, left open the possibility that they were not: "The next best rationalistic explanation of [the phenomena] is that they are fictions, willful or innocent; and that Messrs. Gurney, Myers, and Podmore are victims, partly of the tendency to hoax, but mainly of the false memories and mythopoetic instincts of mankind." James anticipated that "Saturday reviewers will dispose of [*Phantasms of the Living*] in the simplest possible way, by treating the authors as born dupes."[41] This was, for the most part, what happened. In an editorial review in the *Times*, the writer notes that "the undertaking is bold, but the method is not itself unsound," further concluding: "We are not however prepared to affirm that the phenomena adduced in 'Phantasms of the Living' are as yet entitled to this amount of scientific recognition."[42] The reviewer acknowledges the prodigious circulation of such supernatural stories, noting: "Their frequency and persistency may perhaps be regarded as affording a faint presumption in favour of their truth. But the presumption is after all a very faint one. Strict verification is from the nature of the case out of the question." In his review of the book in *Longman's Magazine*, Andrew Lang writes: "It is a most extraordinary and, to a contemplative mind, a most puzzling thing that one never can take Psychical Research seriously."[43] Lang concedes that he believes there is "something in it."[44] However, he is unable to wholeheartedly endorse the book as anything more than as an accumulation of contemporary mythologies. And, in the absence of material evidence, mythology was all the researchers had.

Such doubts as to whether any account of the supernatural could be verified dogged many of the SPR's governing members. In a journal entry from 1887, one year after the publication of *Phantasms of the Living*, Henry Sidgwick, president of the SPR, privately conceded: "I have been facing the fact that I am drifting steadily to the conclusion . . . that we have not, and are never likely to have, empirical evidence of the existence of the individual after death."[45] Indeed, irrefutable proof seemed as elusive as the phantasms of Gurney's study. In his posthumously published autobiography, *Fragments of an Inner Life*, F. W. H. Myers laments the precarious nature of truth in paranormal investigation: "My own career has been a long struggle to seize and hold the actual truth amid illusion and fraud. I have been mocked with many a mirage, caught in many a Sargasso Sea."[46] But if the ensnaring and relentlessly obscure nature of paranormal research was disappointing to Myers, it may have been fatal for Edmund Gurney. Gurney died, some say by his own hand, two years later in a Brighton hotel.[47] Nevertheless, the book continued to attract support and criticism and would eventually inspire imitation.

True Ghost Stories: The Fictional Afterlife of the SPR

In *The Invention of Telepathy*, Roger Luckhurst claims that the critique of the psychical ghost extended beyond debates that were active in the publications of the period. Luckhurst observes that the writer Vernon Lee sought to distinguish her own work from that of the psychical researchers, arguing that for Lee "psychical phantasms were resistant to 'picturesqueness'" and that "'ghosts in the scientific sense' literalized the ghostly associations of the past." Luckhurst further notes that "by 1911, M.R. James had prefaced his tales with a by-then conventional rejection of any 'scheme of psychical theory,' because 'technical terms . . . tend to put the mere ghost story . . . upon a quasi-scientific plane, and call into play faculties quite other than the imaginative.'"[48] Rejection of the "psychical phantasm" was, by the early twentieth century, fairly common among more highbrow authors who sought to elevate the status of the popular ghost story. However, writers, like Henry James, who rejected the "psychical phantasm" were nevertheless well aware of and, some have argued, inspired by the SPR's apparitions. However, there is counterevidence that psychical research in general, and *Phantasms of the Living* in particular, dulled neither readers' fears of nor their fondness for the ghost, nor did it stunt the literary development of subsequent supernatural narratives. If we are to take our evidence from the popular press and the overwhelming response Gurney received from the public at large in his requests for ghost stories, it seems evident that the ghost did not simply disappear, but rather contributed to the formation of a new type literature as it continued to reflect public interest in the new psychical ghost and its new investigator.

If *Phantasms of the Living* failed to convince Victorian readers and critics of the veracity of the truth-telling ghost, the book nevertheless inspired imitation and reanimated the "true ghost story" genre that had been popular during the mid-Victorian period. In 1890, Stead left his post as editor of the *Pall Mall Gazette* to establish the *Review of Reviews*.[49] Stead's journal also covered scientific debates in a manner that was designed to appeal to the lay reader. In the autumn of 1891, the following headline appeared in the review: "Wanted, a Census of Ghosts!" Stead, in an effort to reproduce and ostensibly supplement the research done by Gurney, Podmore, and Myers, began to establish an archive of phantoms of his own. In support of the conclusions reached by the researchers, Stead writes, "The Society has at least succeeded in establishing beyond all gainsaying—first, that apparitions really appear, and secondly, that they are at least as often apparitions of persons living at a distance from the place where the apparition is observed as they are apparitions of those who have died."[50] The bulk of Stead's efforts resulted in *Real Ghost Stories*, an extra published in time for the "ghost season" that allegedly sold the entire run of one hundred thousand copies.[51] By bringing the elevated aims of psychical research to a mass reading public, Stead managed to democratize, and perhaps exploit, this body of knowledge in much the same way that he had democratized scientific knowledge. These new ghost stories were part of Stead's effort to reclaim the ghost from the SPR, an organization he later described as being infected with "a fatal air of sniffiness, as if they were too superior persons to live on the same planet with ordinary folk—some of whom, unlike the Psychical Researchers, sometimes happen to have some psychical gifts of their own."[52] Stead capitalized upon what the researchers disavowed. He gave the ghost back to the "ordinary folk." But scientific trappings replaced the ghost's rattling chains.

After the publication of *Phantasms of the Living*, "psychical ghosts" and "psychical cases" appeared with some regularity in the literary journals of the period.[53] The fictional Victorian ghost, now processed through the critical lens of psychical research, became a durable product of what Virginia Woolf described as the "sudden extension of [the] field of perception."[54] Indeed, the degree to which psychical research had altered popular conceptions of the ghost is made manifest in literary critic Olivia Howard Dunbar's 1905 article concerning the ghost story: "Never before, since spectral feet first crossed a man-made threshold, have ghosts been so squarely, openly, and enthusiastically believed in, so assiduously cultivated, as now. We have raised ghost-lore to the dusty dignity of a science."[55] However, as I have illustrated in this chapter, the ghost showed itself to be an impossible subject for science. Instead the psychical ghost revived the fictional roots of the ghost story itself. What may have limited the psychical ghost's potential as a scientific subject was what made it successful as a story. Nevertheless, this new specter had been significantly redefined by the investigations of the SPR, and this redefinition emerged in countless fictional tales of psychical detectives and ineradicable ghosts. But, as William James had anticipated, such ghosts endured through the "mythopoetic instincts of mankind," issuing forth as tissue-thin substances of the self. These ghostly selves were the new specters of human experience. Indeed, in her reading of Henry James's ghosts, Virginia Woolf notes that these new ghosts were not the chain-laden phantoms of the gothic novel but had "their origin within us."[56] The ghost continues to endure, not as a subject of science, but as the story we tell of ourselves.

Notes

1. Virginia Woolf, *Essays of Virginia Woolf*, vol. 2, *1912–1918*, ed. Andrew McNeillie (London: Harcourt Press, 1986), 219. For more on Woolf's response to, and interest in, psychical research and psychology, see George M. Johnson, "'The Spirit of the Age': Virginia Woolf's Response to Second Wave Psychology," *Twentieth Century Literature* 40, no. 2 (1994): 139–64.

2. Edmund Gurney, Frederic W. H. Myers, and Frank Podmore, *Phantasms of the Living*, 2 vols. (London: Trübner, 1886).

3. Srdjan Smajic, *Ghost-Seers, Detectives, and Spiritualists: Theories of Vision in Victorian Literature and Science* (Cambridge: Cambridge University Press, 2010). Pamela Thurschwell mentions *Phantasms of the Living* in her *Literature, Technology and Magical Thinking, 1880–1920* (Cambridge: Cambridge University Press, 2001), as does Marina Warner in *Phantasmagoria* (Oxford: Oxford University Press, 2006). However, Warner misidentifies its date of publication (241). In her "Psychic Stories," Lisa Brocklebank also refers to the text parenthetically, but focuses primarily upon reading and

F. W. H. Myers's theories of telepathy (*Victorian Studies* 48, no 2 [2006]: 233–39). For a readable and thoroughly entertaining introduction to the history of the SPR and *Phantasms of the Living*, see Deborah Blum's *Ghost Hunters: William James and the Hunt for Scientific Proof of Life after Death* (New York: Penguin, 2006).

4. Roger Luckhurst, *The Invention of Telepathy* (Oxford: Oxford University Press, 2002), 72.

5. Shane McCorristine, *Specters of the Self: Thinking about Ghost and Ghost-Seeing in England, 1750–1920* (Cambridge: Cambridge University Press, 2010), 5.

6. Renée Haynes, *The Society for Psychical Research, 1882–1982: A History* (London: Macdonald, 1982), 12–13. The SPR was not the first Victorian organization to investigate supernatural phenomena. In 1869, two years after it was established, the London Dialectical Society began study of spiritualist phenomena. Like the SPR, the London Dialectical Society included prominent intellectuals within its ranks. Political activist Charles Bradlaugh and the naturalist Alfred Russel Wallace were both members. However, "Darwin's Bulldog," T. H. Huxley, and the skeptical philosopher George Henry Lewes notably refused when both were invited to join. In 1875, Edward William Cox established the Psychological Society. The spiritualist medium William Stainton Moses and C. C. Massey, a noted theosophist, were among its first members. This society's investigations were not quite as rigorous as those conducted by the SPR. The organization nevertheless was among the first to attempt rational study of spirit world but eventually dissolved after Cox's death in 1879.

7. Bart Schultz, *Henry Sidgwick, Eye of the Universe: An Intellectual Biography* (Cambridge: Cambridge University Press, 2004), 276.

8. F. W. H. Myers, *Fragments of an Inner Life* (London: Society for Psychical Research, 1961), 14.

9. Haynes, *Society for Psychical Research*, 12–13.

10. Alan Gauld, *The Founders of Psychical Research* (London: Routledge & Kegan Paul, 1968), 153.

11. Gurney, Myers, and Podmore, *Phantasms of the Living*, 1:xlvi n. 1.

12. Ibid., lxiii.

13. Luckhurst, *The Invention of Telepathy*, 43.

14. Gurney also published requests in multiple issues of *Trewman's Exeter Flying Post*, the *Hampshire Telegraph*, and the *Northern Echo*. In these requests, Gurney asked potential contributors whether or not they "when in good health and completely awake had a distinct impression of seeing or being touched by a human being, or of hearing a voice or sound which suggested a human presence, when no one was there" or if they had ever experienced "a dream of the death of some person known to you (about whom you were not anxious at the time), which dream you marked as an exceptionally vivid one, and of which the distressing impression lasted for as long as an hour after you rose in the morning." "Yes or No," *Standard*, December 17, 1883. The popular press did its share in helping Gurney in his task. In an article from December 17, 1883, a writer from the *Pall Mall Gazette* suggests that the reader "who is going down into the country for the holiday . . . [take] a ghost census of his fellow-passengers" and lightheartedly thanks the society for this "capital Christmas suggestion." "Occasional Notes," *Pall Mall Gazette*, December 17, 1883.

15. Gurney, Myers, and Podmore, *Phantasms of the Living*, 1:xlix.

16. "Second Report of the Committee on Haunted Houses," *Proceedings of the Society for Psychical Research*, vol. 2 (London: Trübner, 1884), 139. http://hdl.handle.net/2027/mdp.39015013709335.

17. Ibid., 140.

18. Andrew Lang, *Cock Lane and Common Sense* (London: Longmans, Green, 1894), 138.

19. The British novelist Mrs. H. D. Everett (Theo Douglas), an author who H. P. Lovecraft claimed "reache[d] singular heights of spiritual terror" in late Victorian and Edwardian fiction, casts the "true" ghost in similar terms. H. P. Lovecraft, "Supernatural Horror in Literature," in *At the Mountains of Madness: The Definitive Edition*, ed. China Mieville (New York: Modern Library, 2005), 154. In her story "Anne's Little Ghost," the narrator notes that "the true ghost story never has any point and the faked one dare not leave it out. See H. D. Everett, *The Crimson Blind and Other Stories* (Hertfordshire: Wordsworth Editions, 2006), 122.

20. Gurney, Myers, and Podmore, *Phantasms of the Living*, 1:435.

21. Ibid.

22. Ibid., 2:462.

23. Ibid., 515.

24. Ibid.

25. A. Taylor Innes, "Where Are the Letters? A Cross-Examination of Certain Phantasms," *Nineteenth Century* 126 (August 1887): 174.

26. Catherine Crowe, *The Night Side of Nature or Ghosts and Ghost Seers* (London: Newby, Mortimer, and Cavendish, 1848); Robert Dale Owen, *Footfalls on the Boundary of Another World with Narrative Illustrations* (Philadelphia: J.B. Lippincott, 1860).

27. W. T. Stead, ed., *Real Ghost Stories: A Record of Authentic Apparitions* (London: Publishing Office of the Review of Reviews, 1891); Andrew Lang, *The Book of Dreams and Ghosts* (London: Longmans, Green, 1897); Elliott O'Donnell, *Byways of Ghost Land* (London: W. Rider and Sons, 1911). Lang often cites *Phantasms of the Living* in his study and apparently used the book as a source text. He also includes this study as an example, along with Catherine Crowe's *Night-Side of Nature*, of ghostly literature in his book *The Origin of Religion*.

28. Trevor H. Hall, *The Strange Case of Edmund Gurney* (London: Duckworth, 1964), 72.

29. In 1923, Lodge was asked to share his opinion of the modern ghost story for the British literary periodical *Bookman*. Lodge had lost his son in the war eight years earlier and had subsequently published *Raymond*, an enormously popular book that detailed the author's communications with the dead. By this time Lodge had changed his mind about the veridical phantasm: "What the public is really interested in is the amount of underlying truth, and the meaning that may be involved, in supernormal experiences. To arrive at sound conclusions demands careful and continued and unbiased study; the concoction of imaginary narratives is useless to that end, and is not what the public really wants." "Dreams, Ghosts and Fairies," *Bookman* 65, no. 387 (December 1923): 146.

30. Andrew Lang, "Ghosts Up to Date," *Blackwood's Edinburgh Magazine* 155, no. 939 (January 1894): 47.

31. "No Vacancy: Some Very Dull Ghosts," unsigned review of *Phantasms of the Living*, by Edmund Gurney, Frederic W. H. Myers, and Frank Podmore, *Daily News*, November 5, 1886.

32. Gurney, Myers, and Podmore, *Phantasms of the Living*, 1:186.

33. "Phantasms of the Living: A New Apology for Ghosts," unsigned review of *Phantasms of the Living*, by Edmund Gurney, Frederic W. H. Myers, and Frank Podmore, *Leeds Mercury*, March 26, 1887.

34. "Spookical Research," *Saturday Review of Politics, Literature, Science and Art* 62, no. 1620 (November 1886): 648–51.

35. "Psychical Research," *Pall Mall Gazette*, October 21, 1882. The *Review*'s humorous coinage "spookical" reflects the attitude that many late Victorian critics bore toward the SPR's investigations. For these critics, ghosts—whether crisis apparitions or veridical hallucinations—were ultimately little more than whimsical spirits that belonged in the Christmas annuals. Because the SPR endeavored to study "ghosts," such critics regarded the society's investigations with a mixture of disbelief and derision.

36. Because of a warehouse fire, Trübner and Co. were forced to delay the publication of *Phantasms of the Living* until 1886. However, the popular press reproduced the book's cases from November 1885 to January 1886, thus satisfying a public readership accustomed to seasonal apparitions. See "Some of Mr. Gurney's Ghost Stories," *Pall Mall Gazette*, December 16, 1886; "'A Ghost Sent to Prison . . . ,'" *Standard*, December 25, 1886; and "Some Modern Ghost Stories," *Derby Mercury*, December 29, 1886.

37. "No Vacancy: Some Very Dull Ghosts," *Daily News*, November 5, 1886.

38. George Bernard Shaw, *Bernard Shaw's Book Reviews*, ed. Brian Tyson (University Park: Pennsylvania State University Press, 1991), 2:224.

39. George Bernard Shaw received a copy of the book and in a November 1886 review flatly denied the plausibility of Gurney's phantoms, stating that "the existence of a liar is more probable than the existence of a ghost" (ibid., 219). As for the SPR's witnesses, Shaw notes that "the great majority are so ignorantly convinced that seeing is believing, so little aware that the evidence of their senses requires highly skilled interpretation before it can exert weight in the balance of science, that they are the last person who can be depended upon to give a trustworthy account of what has actually passed in their presence" (219). Not all reviewers were as critical of *Phantasms of the Living* as Shaw and his colleagues. In his review of the book in the January 27, 1887, issue of *Nature*, the British psychologist C. Lloyd Morgan noted that "normal psychology has much to learn from experiments on supernormal and abnormal 'subjects'" and concluded by suggesting that if Gurney's and Myers's work "throw[s] light upon "hidden mysteries, which are none the less realities, of the human mind, their labor will, in my opinion, not have been in vain." "Supernormal Psychology," *Nature*, January 1887, 292, http://hdl.handle.net/2027/coo.31924066894019.

40. "Some Modern Ghost Stories," *Derby Mercury*, December 29, 1886.

41. William James, review of *Phantasms of the Living* by Edmund Gurney, Frederic W. H. Myers, and Frank Podmore, *Science* 9, no. 205 (January 1887): 19, http://www.jstor.org/stable/1761001.

42. "How Many Senses Have You?," unsigned review of *Phantasms of the Living* by Edmund Gurney, Frederic W. H. Myers, and Frank Podmore, *Times*, October 30, 1886.

43. Andrew Lang, "At the Sign of the Ship," review of *Phantasms of the Living* by Edmund Gurney, Frederic W. H. Myers, and Frank Podmore *Longman's Magazine* 9, no. 51 (January 1887): 330.

44. Ibid., 332.

45. Janet Oppenheim, *The Other World: Spiritualism and Psychical Research in England, 1850–1914* (Cambridge: Cambridge University Press, 1985), 117.

46. Myers, *Fragments*, 40.

47. Colin Wilson argues that the fraudulent nature of paranormal investigation beyond *Phantasms of the Living* led Edmund Gurney to an early death, alleging: "Gur-

ney committed suicide in a Brighton hotel in 1888 when he discovered that certain trusted mediums were tricksters." *The Occult* (London: Watkins, 2004), 493. Trevor Hall, in his biography of Gurney, seems to concur with Wilson's verdict. Indeed the primary focus of Hall's book concerns the alleged deceit that two men, Douglas Blackburn and George Albert Smith, perpetrated on the Society during repeated demonstrations of thought-transference in the 1880s. In Hall's view, the revelation of this deceit impacted Gurney to a large degree since George Smith was also Gurney's friend and private secretary. But Hall also places the responsibility of Gurney's death at the feet of his colleagues Henry Sidgwick and F. W. H. Myers. Hall substantiates this claim by citing an observation made by a friend of the Gurneys, Lady Constance Battersea. In her book *Reminiscences*, Battersea notes that Edmund Gurney's mental condition was in decline during the time preceding his death. She also, rather dramatically, avers about Gurney that, "then being infected by the enthusiasm of his friends, Mr. Myers and Mr. Sidgwick, he gave himself unreservedly to the problems of psychical research, that perplexing and elusive subject, and devoted his time, his pen, and alas! his strength, to that which seemed always evading his grasp." *Reminiscences* (London: Macmillan, 1922), 40.

48. Luckhurst, *The Invention of Telepathy*, 187.

49. J. O. Baylen, "W.T. Stead as Publisher and Editor of the 'Review of Reviews,'" *Victorian Periodicals Review* 12, no. 2 (1979): 70–71, http://www.jstor.org/stable/20081887.

50. W. T. Stead, "Wanted, a Census of Ghosts!" *Review of Reviews* 4, no. 21 (September 1891): 257.

51. Luckhurst, *The Invention of Telepathy*, 120.

52. W. T. Stead, ed., *Borderland: A Quarterly Review and Index*, vol. 2 (London: Mowbray House, 1895), 346, http://hdl.handle.net/2027/mdp.39015074624928.

53. For example, see B. F. Cresswell's story "A Psychical Experiment," *Belgravia: A London Magazine* 64 (February 1888): 26; "A House of Horrors," *Belgravia* 95 (January 1898): 1; MBC, "A Western Mystery," *Belgravia* (June 1892): 158; and George Makgill, "A Case for the Psychical Society," *MacMillan's Magazine* 1, no. 6 (April 1906): 471.

54. Woolf, *Essays*, 219.

55. Olivia Howard Dunbar, "The Decay of the Ghost in Fiction," *Dial* 38 (1905): 378, http://hdl.handle.net/2027/inu.32000000678435.

56. Woolf, *Essays*, 324.

CHAPTER 13

The Energy of Belief

The Unseen Universe *and the Spirit of Thermodynamics*

Tamara Ketabgian

In the latter half of the Victorian period, new theories of energy predicted a future as bleak as it was exciting. Promising both the end of the world and its ambitious mastery through theory, this vision followed the contemporary formulation of the two laws of thermodynamics by scientists Hermann von Helmholtz, Rudolf Clausius, and William Thomson (later Lord Kelvin). The first law—of conservation—portrayed a redemptive world of unified, indestructible, and ever-convertible energy, often likened to the eternal power of the divine. The second law, however, was nothing short of apocalyptic, foretelling the path of all closed energy systems to entropy—to waste, disorder, and irretrievably dissipated heat, with the end result of cosmic extinction.[1] Critics Gillian Beer, Bruce Clarke, and George Levine have all eloquently addressed this "degenerative vision" of a finite, fallen universe—a godless world haunted by "terrors" of "death irrecuperable."[2] The sun and earth would most certainly perish, as Thomson calculated in 1862; and, according to H. G. Wells, thirty million years hence, the sole remnants of life might well be only a few giant crabs, lichens, and liverworts.[3] This chapter examines a particularly fanciful and controversial response to Thomson's grim thermodynamic scene: *The Unseen Universe, or Physical Speculations on a Future State* (1875), a work of polytheistic cosmic science that reimagines energy physics as a site of transformative spiritual consolation. In this surprisingly popular treatise, energy serves as a metaphor for belief itself, imagined as a physical force testing the bounds of multiple, invisibly linked worlds.

Coauthored by two established professors of physics, Balfour Stewart (1828–1887) and Peter Guthrie Tait (1831–1901), *The Unseen Universe* redeems the world's wasted energy as evidence of a possible spiritual afterlife, located in an immortal and invisible beyond. Yet, above all, Stewart and Tait's text is a determined exercise of rhetoric, in which the work of energy and persuasion are one and the same. Blending Christian apologetics with astronomy, thermodynamics, and a rather partial survey of world religion, their treatise uses the vast dimensions of modern physics—of cosmic breadth and finitude—to evoke a new affective landscape of religious conviction, speculation, and influence. Stewart and Tait thus revise the traditional clockwork view of natural theology to yield a more dynamic vision of the universe as a series of heat engines nested within the luminiferous ether.

While received enthusiastically by mystics and spiritualists, *The Unseen Universe* spurred widespread scientific criticism for its speculative fictions and "philosophical abstractions."[4] According to critic William Kingdon Clifford (1845–79), this text merely replaces one form of primitive faith—"the good old gods of our race—sun, sky, thunder, and beauty"—with another—the scientistic natural concepts of "substance, energy, and life, under the patronage respectively of the persons of the Christian Trinity" (792). However, as Stewart and Tait argue in the case of vortex atoms, ether, and—of course—energy, many of these concepts are invisible to empirical human observation, and therefore must rely on imaginary representation, speculation, and, ultimately, faith. Through such models and analogies, adapted from natural theology, *The Unseen Universe* seeks both to accommodate the spiritual limitations of its readers and to critique agnostic, materialist scientists for their reliance on similarly abstract fictions. Neither fully provable nor refutable, this thermodynamic fantasy deeply questions the limits of knowledge, the legitimacy of evidence, the purpose of analogy, and the relation between modern science and modern belief.

Beyond its contemporary renown, Stewart and Tait's treatise led a vigorous afterlife in the years following its publication. The conclusion of this chapter explores *The Unseen Universe*'s forgotten legacy: its formative influence on both popular occult practice and still-urgent philosophical debate on scientific method and the psychology of belief. *The Unseen Universe* inspired enthusiastic citation by Helena Petrovna Blavatsky (1831–91), founder of the Theosophical Society and author of *Isis Unveiled* (1877), an exhaustive account of mystical energy transfer. Yet, aside from Blavatsky, *The Unseen Universe* was the crucial premise for philosopher William James's (1842–1910) well-known exchange with Clifford on the ethics of faith—a topic that even now informs public discourse surrounding truth claims, scientific authority, and what James later termed "The Will to Believe" (1896). Still shaping how we argue about the unseen, this strange treatise allies scientific belief with the most ethereal of subjects: energy—its loss, its transfer, and the fantasy of its eternal return.

Invisible Worlds and Entropic Redemption

The Unseen Universe sold rapidly and consistently, appearing in no less than fourteen editions in thirteen years. Although its first three editions were published anonymously, its authorship was widely known within scientific circles.[5] In the text's first preface, Stewart and Tait claim their central object is "to show that the presumed incompatibility of Science and Religion does not exist."[6] Poised at the unsteady juncture between professional science and natural theology, their treatise insists on a method "absolutely driven by scientific principles."[7] Its actual procedure is more hybrid and eccentric, blending thermodynamic models of heaven with biblical citations that arguably also provide "trustworthy communication" of the universe's "intelligent agency" (223–24). Even so, Stewart and Tait were hardly renegades from the evolving British scientific establishment. Solving the problem of entropy symbolically, with the creation of new, energy-filled worlds, *The Unseen Universe* reflects this establishment's divided stance toward theistic science and its quasi-spiritual vestiges in natural law.

As theistic physicists from the "North British" energy school,[8] Stewart and Tait held influential roles within mid- and late-Victorian natural philosophy. Both men came of age during the later years of this period's "undisciplined culture"[9] and were marked by its broad, predisciplinary mingling of art, science, and speculative theology. Stewart attended Edinburgh University, served as director of Kew Observatory, and then occupied a chair in physics at Owens College, Manchester, where he earned the Rumford Medal for his research on radiant heat in 1868.[10] A senior wrangler in mathematics at Cambridge, Tait became professor at Queen's College, Belfast, subsequently at Edinburgh University, and was recognized for his collaboration with Thomson on both thermodynamic theories of heat dissipation and their coauthored *Treatise on Natural Philosophy* (1867).[11] Like Thomson and other theistic colleagues, Stewart and Tait opposed the secular materialist worldview that John Tyndall so dramatically promoted in his 1874 presidential "Belfast Address" to the British Association for the Advancement of Science. *The Unseen Universe* pointedly responds to Tyndall's call to "wrest from theology, the entire domain of cosmological theory."[12] Tyndall claimed that life was nothing more than a "purely physical condition,"[13] a simple conversion of energy that—as Graeme Gooday notes—distressed Stewart and Tait precisely because it left "no scope for the exercise of the soul or free will."[14] To counter this restrictive determinism, their text offers an alternate "philosophy of nature,"[15] whose Christianized physics supports the prospect of immortality in a "spiritual universe" that is "full of life and intelligence" (5).

Despite their staunch antimaterialism, Stewart and Tait share many common rhetorical strategies with their agnostic scientific critics, including practices from natural theology, a tradition of religious knowledge based both on the direct study of nature and, frequently, on a "discourse of design."[16] For while we might assume natural theology lost favor after the rise of aleatory Darwinian theory and thermodynamics, the scientific writings of many late-Victorian agnostics still retain powerful vestiges of faith. As historian Bernard Lightman has shown, agnostics such as Tyndall produced a "new natural theology" through their emphasis on empirically observed natural law, order, and beauty.[17] Similarly, many critics, such as John Hedley Brooke and Jonathan Topham, have traced the emergence of an alternate "theology of nature"[18] in the period's professional scientific language, which extols the grandeur of the physical world. Even natural law yielded potentially religious effects, as Barri Gold recognizes in the "grand unified theories" of science that dominated the nineteenth century.[19] Gold argues that the "development of thermodynamic concepts among physicists" supported powerful forms of "faith" and "faithlike" conviction, by reconceiving dissipated energy not as "loss" but as meaningful change.[20] Like these implicitly spiritual scientists, Stewart and Tait pose their own redemptive "grand theory" of heaven as a site of energy recycling, treating entropy not as a tragedy but as an opportunity for renewed belief.

Aside from this shared spiritual rhetoric, however, Stewart and Tait approach natural phenomena with a speculative openness that departs dramatically from the methods of their materialist peers. Throughout *The Unseen Universe*, they dwell on the massive spilt energy of our current "visible" world:

> All but a very small portion of the sun's heat goes day by day into what we call empty space. . . . Could anything be more perplexing than this seemingly prodigal expenditure of the very life and essence of our system? That all but a petty fraction of this vast store of high-class energy should be doing nothing but travelling outwards in space at the rate of 188,000 miles per second is hardly conceivable, especially when the result of it is the inevitable destruction of the visible universe. (197)

Instead of conversion into other productive uses, this lost power is a "perplexing" dead end in the form of stagnant and chaotically dispersed heat. Starkly apocalyptic, this wasted energy ultimately signals an "earth unfit for habitation" (196) and a universe fated to end. Yet, as a symbol, this scene of entropy also reveals telling distinctions between secular science and *The Unseen Universe*'s more byzantine project of spiritual interpretation. Whereas materialists marveled at the world's waste as an affecting mark of its finitude, Stewart and Tait viewed this excess as a physical—and metaphysical—sign for which science must account. Unlike agnostics such as Tyndall, who treated physical laws and objects as sufficient consolations in and of themselves, Stewart and Tait regard these spectacles as expansive spiritual traces, betokening invisible correspondences beyond the reach of rational inquiry. Ultimately, their treatise suggests that privileged meaning does not lie in material forms, but in what these forms express in other transcendent registers.

In Stewart and Tait's symbolic logic, waste paradoxically asserts the presence of its own hidden corrective: heaven. Lost power serves as evidence of a "state of intimate mutual relation" between "visible and invisible" worlds (248), thus signaling the existence of a fantastical "*paraspace*"[21] an alternate realm that doubles as a thermodynamic heat sink. For Stewart and Tait, this heaven-like site is "connected by bonds of energy with the visible universe." This invisible world both "receiv[es]" and "transform[s]" our spent power, through acts of gradual transfer and absorption (199). Supporting new forms of spiritual insight, our leaky universe is invisibly linked to another, through redemptive bonds, "pulses," "rents," and "cracks" (198, 221).

More than proof of an alternate world, this pervasive energy transfer also reveals the immortal, miraculous, and infinite nature of heaven. In *The Unseen Universe*, death fuels immortal life, and the decline of our world will sustain the vitality of the next. Reversing the course of entropy, the resulting "invisible universe" will be "full of energy when the present universe is defunct" (200). The immortality of the human soul results from this same exchange, which will preserve the material "motions which accompany thought" (199) and, with them, both the mind's "power of action in the present" and its "hold upon the past" through memory (200). For all the atoms that form matter, Stewart and Tait's site is both the final end and original source of our present world's power. As in a machine or other closed thermodynamic system, their unseen universe serves as the necessary "material antecedent" for "each transformation of energy" (184). Stewart and Tait thus follow popular nineteenth-century models of entropic creation, which, according to critic Helge Kragh, treat our world as a "finite-age, created universe" that "must have had a beginning."[22] Even miracles stem from "transmutations of energy from the one universe into the other," governed by a "principle of Continuity" (vii)[23] that supports "an endless chain of events" (257–58).

While Stewart and Tait remain vague on the actual form and content of heaven, they base their vision on analogies largely from fluid mechanics. At first, both authors speculate on our world's link to a single alternate universe, which is revealed through the irregular surface film of a greater ethereal fluid:

> We may suppose our (essentially three-dimensional) matter to be the mere skin or boundary of an Unseen [universe] whose matter has *four* dimensions. And, just as there is a peculiar molecular difference between the surface-film and the rest of a mass of liquid—wherever such a surface-film exists, even in the smallest air-bubble—so the matter of our present universe may be regarded as produced by mere rents or cracks in that of the Unseen. (221)

This fluid model is in turn succeeded by a more rarefied field of many bubble-worlds, forming "four-dimension boundaries of the five-dimensional matter of a higher Unseen, and so on" (221), in an image that anticipates postmodern accounts of the multiverse in quantum physics.[24] Stewart and Tait thus imagine "an infinite series of Universes, each depending on another" and sharing among them both "an infinite store of energy" (222) and a deity of sorts—an "intelligent developing agency" (221). Compared to bubbles, heat engines, and other linked yet contained systems, these multiple worlds arise within the fluid, invisible medium of the ether, another abstract scientific model that this essay will subsequently explore. Throughout *The Unseen Universe,* Stewart and Tait openly concede the obscurity of such models—and suggest that human limitations are to blame: our inability to grasp refined spiritual phenomena "explain[s] . . . how it is that so very little of the nature of the definite description of the Unseen is given" (221–22).

The Unseen Universe is a strange and often unaccountable work, claiming scientific authority while it defends its own obscurantism as necessary and inevitable. Yet, more than a spiritual apologetic or compromise formation, this text actively employs concepts from Victorian energy physics to support its unusual spiritual speculations. Through inventive analogy, Stewart and Tait rewrite the pivotal drama of fin de siècle physics—what philosopher Michel Serres has termed the birth of an "irreversible and irrevocable" view of "thermodynamic time."[25] In this new image of a "closed-isolated" world, "energy dissipates," "entropy increases," and "time is endowed with a direction. It is irreversible and drifts from order to disorder," and, finally, to death.[26] While other Victorians mourned this finite world of entropic closure, Stewart and Tait seek to reopen it with their own grand theory of a multiverse fueled by infinite energy and belief.

Natural Theology and the Energy of Persuasion

The Unseen Universe's emphasis on energy is deeply intertwined with its goals and structure as a work of natural theology. As an updated apologia of thermodynamics as divine law, this text reads the Book of Nature as a series of interlinked engines, defined both by the mechanical powers of life and the vital powers of the machine.[27] Following physicist Sadi Carnot (1796–1832),[28] Stewart and Tait compare the functions of the universe to those of a steam motor, defined by its conversion of force, its containment of pressure, and its enclosed chambers of hot and cold—all fueled by the heat of the sun. In their resulting geography of spiritual revelation, energy is harnessed and recycled in an extensive network of closed—and potentially open—systems. What is more, in *The Unseen Universe* this same irresistible energy is also that of persuasion, working toward the spiritual and thermodynamic conversion of readers. Stewart and Tait thus use a blend of mechanical and theological rhetoric—of analogy, accommodation, and divine design—to portray belief as a relentless force driven to transcend limits.

Treating their universe as a collection of subtly crafted devices, Stewart and Tait adopt William Paley's still-popular brand of late eighteenth-century natural theology, which compares natural objects to watches that display skill, artistry, and intrinsically mechanical design. As Paley stresses in his treatise *Natural Theology* (1802), the study of these objects will show that "there cannot be design without a designer, contrivance without a contriver, order without choice."[29] For Paley, this felt experience of "contrivance" is our closest access to the divine, offering a path to spiritual faith and wisdom through technical objects, narratives, and systems.[30] Grounded in a common discourse of design, this language of mechanical immersion is inseparable from natural theology and its allied narratives of divine intention and intelligence.

As figurative aids for spiritual persuasion, these mechanical forms follow the theological doctrine of "accommodation," which argues that "God adapts himself to human capacity in his revelation,"[31] adjusting his "infinite mysteries" to our limited minds, souls, and senses.[32] As Paley notes, "It is only by the display of contrivance, that the existence, the agency, the wisdom of the Deity, *could* be testified to his rational creatures" (38–39). He defines these contrivances as God's rhetorical exercise for humanity's benefit,[33] addressing our restricted "faculties [as] formed at present" (40). Indeed, Paley asks, otherwise "why resort to contrivance, where power is omnipotent?" (38–39). Life and the natural world thus serve as analogic, mechanical models of divine wisdom, purposely designed to aid people in their religious explorations. As viewed by Paley, analogy both spurs rational analysis and suggests an alternate spiritual realm, accessible only through "contrivance" and accommodation.

The Unseen Universe adopts a similar interpretive process, one rooted in the phenomenal experience of complex and often inaccessible technical objects—namely, the world's "vast heat-engine" (126), which is in turn populated by other engines both literal and figurative, both "animate" and "inanimate" (182). Invoking a Paleyean discourse of design, Stewart and Tait emphasize the delicacy, complexity, and obscurity of our world's "animate" machines. These living systems support more "sudden and violent . . . transmutation[s] of energy," much like a "rifle at full-cock, with a delicate hair-trigger . . . where the slightest touch from without may bring about the explosion of the gunpowder, and the propulsion of the ball with a very great velocity" (183). According to our authors, a similar process may occur in human beings, where "a very small and obscure transmutation of energy in the mysterious brain chamber may determine some very violent motion" (185). The resulting actions may seem abrupt or irregular to our limited perceptions as viewers. However, as both physicists conclude, we are simply unaware of their delicate origins and the greater divine and mechanical causes that govern such unpredictable acts.[34]

While *The Unseen Universe* stresses life's obscure complexity for human viewers, it dwells most pointedly on the spiritual knowledge of its fallible and receptive readers. In general, Stewart and Tait suggest that *all* people require divine aid to amplify their restricted vision: like ants, human beings view the uprooting of their hill as a process "mysteriously perplexing, far transcending their experience, but *we* know, very well that the whole affair happens without any breach of continuity of the laws of the universe" (248). Yet, while all mankind is arguably antlike in this scenario, the authors address a more specific audience—the ambiguous and presumably more enlightened "we" of the passage. Stewart and Tait write neither for confident religious believers, nor for extreme scientific materialists (70–71), but for "honest doubters" (202), who "see strong grounds for belie[f]" yet have "deeply studied the scientific objections and do not well see how to surmount them" (71). This ideal audience is vital to the *Unseen Universe*'s success as an immersive experience of spiritual belief and interpretation.

Relying on the receptivity of its readers, this treatise resembles spiritualist experiments later conducted by the Society for Psychical Research and even analyzed by Stewart himself as a SPR council member and president seeking to document the empirical conditions for observing "psychic force" and telepathy.[35] Spiritualists commonly faulted skeptical materialist observers for disrupting the delicate environment required for experimental success, as when Tyndall undermined the necessary "mental and physical passivity" for a séance in 1864.[36] For such mediums, as for Stewart and Tait, the spiritual disposition of observers would crucially shape the success and authority of their efforts at persuasion. *The Unseen Universe* compares this task of reception to "a great steamer . . . carrying two sets of passengers": religious thinkers on deck, concerned with the greater purpose and direction of their voyage, and scientific thinkers "remaining below," who question how the steamer's engines actually work (25). With its "honest doubters," Stewart and Tait's ideal audience ranges *both* above and below deck, dwelling both on broader visionary theory and on more empirical practices, in a stance that combines abstract and material ways of seeing.

As Stewart and Tait show in the hybrid viewpoint of their reader-passengers, mechanical analogy serves as the ultimate vehicle for accommodation, inviting not only detailed technical interpretation but also speculation on realms beyond human analysis. Defined by philosophers and theologians as a specialized technology of perception, employing cognitive "correction and qualification,"[37] analogy uses comparison to bridge the gap between our ordinary sense of language and a more speculative sense where such language might apply to God. In his review of *The Unseen Universe,* William James notes precisely this process of spiritual immersion and extrapolation in readers who—through analogy—use their own "faith or fancy [to] . . . fill out [the] details" of Stewart and Tait's models of mechanical natural law and "continuity."[38] Here analogy supports the accommodation of limited human faculties, as a technical and rhetorical "contrivance" intended to aid spiritual exploration.

As a form of religious argument, analogy has long been faulted for falsely comparing (and conflating) its objects with a divine referent. Stewart and Tait recognize the figure's fictive aspects and acknowledge their role in supporting similar forms of linguistic immersion and approximation. They concede, "We are to some extent constructors, and find analogies in nature which seem to us to throw light upon the doctrines of Christianity" (vii). As Stewart and Tait suggest, they seek to estimate and *simulate* scientific proof rather than to provide such evidence according to accepted empirical standards: "Although our evidence from analogy may not amount to proof, it is very strong. What we have done is to show that a future state is possible, and to demolish any so-called scientific objection that might be raised against it. Evidence in favour of the doctrine is not derived from us" (211). Aiming only to speculate on possibilities, they do not intend to prove the affirmative presence of the unseen universe but rather that it is *not impossible*: "There is nothing . . . to lead us to suppose that life is impossible after death" (10). Similarly, they argue, "We cannot deny the possibility of a future life" (6). As in many works of natural theology, this language of double negatives, conditionality, passivity, and constraint is also one of affective force—of emotional energy and the power of belief. Stewart and Tait "*feel constrained* to believe"; they "*are forced* by a purely scientific process to recognize the existence of an Unseen Universe" (6–7).[39] Their evidence from analogy is "strong" (211), affecting, and "absolutely *driven by* scientific principles" (5).[40] Ideally receptive, like the members of a spiritualist séance,

Stewart and Tait's readers are materially moved toward belief, through an unseen transfer of affective energy between the abstract tenor and technical vehicle of analogy. In effect, this transfer stages—both psychically and analogically—the same conversion and release of power that occurs between Stewart and Tait's invisible worlds.

The Unseen Universe locates this energy of belief in a cosmic geography of breadth and finitude—of barriers, breaks, and "grand avenue[s]" (16). For Stewart and Tait, our current visible universe is "a limited area bounded by an impenetrable wall, which, if we could only pierce it, would admit us at once into the presence of the Eternal" (96). This confined "area" evokes both the limits and possibilities of the period's newly evolving disciplines, including fields (physics) formerly housed in the more expansive realm of natural philosophy. Here the scientist seeks "to clear a space . . . from which all mystery shall be driven away" and where "nothing [exists] but matter and energy subject to certain definite laws which he can comprehend" (237). Yet, as Stewart and Tait suggest, this "little clearing" (viii) cannot address the greater spiritual mysteries "of the soul's domicile," "of life and intelligence," and "of God" (237). Grounded in empiricism, this limited "circle" (237) bars any speculation beyond its own visible and materially verifiable phenomena. Accordingly, the unseen "avenue" between physics and theology "has been walled up and ticketed with '*[n]o road this way*,' professedly alike in the name of science at the one end, and in the name of the religion at the other" (272).

Despite these setbacks, *The Unseen Universe* still urges us to dwell closely on the limits of our small circle of knowledge. If we wish to grasp "the properties of the unknown lying beyond or at the boundary of our little 'clearing,'" we must "go up to it and examine it often, with long continued labour, under great difficulties" (viii). For, as Stewart and Tait suggest, these sites of closure nonetheless contain significant breaks and anomalies, which "are in reality so many partially concealed avenues leading up to the unseen" (250). Like our world's "lost" energy, these breaks reveal a miraculous "universe within a universe, a portion cut off by an insurmountable barrier from the domain of scientific inquiry" (90). Moreover, the actual closure of this "barrier" even more powerfully signals the presence of the "Eternal" (96), in forms of vision beyond the literal and material. Throughout *The Unseen Universe*, these borders sustain intensity of faith through speculation born of confinement. Their very limits allow for the possibility of an infinite beyond, reconciling our universe's closed world with the open realm of divinity.

Confining one space in order to leave another unbound, *The Unseen Universe* performs the spatial and rhetorical equivalent of a double negative: its barriers *cannot negate*—and thus arguably affirm—the prospect of a mystical site that transcends scientific measure and method. For Stewart and Tait's receptive readers, these boundary worlds restage the process of empirical observation as a capacious field of spiritual and mechanical extrapolation. Through these scientific fictions, analogies, and accommodations, *The Unseen Universe* enacts nothing less than a fantasy of breaching our closed world and reversing the relentless path of entropy and time. As a visionary blend of natural theology and thermodynamics, this text recasts systemic closure as a transformative avenue of openness and abundance.

Imaginary Science and the Ether

Testing the limits of different worlds, disciplines, and beliefs, *The Unseen Universe* uses boundary play to simulate an alternate realm of hypothetical scientific law. In effect, Stewart and Tait pose a form of "imaginary science," to invoke theorist Istvan Csicsery-Ronay's view of science fiction as a speculative genre that employs "an *image* of science, a poetic illusion disguising its illusionary status."[41] Like science fiction and, arguably, many works of natural theology, *The Unseen Universe* treats science as an immersive symbolic, speculative, and technically specialized mode of representation. Yet, as their successive prefaces and revisions show, Stewart and Tait are also clearly sensitive to attacks on their "imaginary" enterprise. Throughout *The Unseen Universe*, Stewart and Tait argue that Clifford's charge—of abstract, fictive representation—could also be leveled toward all established science. For, as they repeatedly ask, when is science *not* fiction? How might we distinguish science from "imaginary science"—from speculative fancy or false analogy? These questions arise most pointedly in the case of atoms and the ether, as two conceptual abstractions that ground both mainstream late Victorian physics and *The Unseen Universe*'s more eccentric reflections.

Stewart and Tait openly recognize the provisional, statistical, and analogical role of scientific law and representation, especially in the field of thermodynamics. They argue that analogy and speculation *best* capture the shifting picture of this world, with its unexpected breaks and deviations. Promoting a more flexible approach toward scientific law as a figural representation, they argue for

> a more complete and continuous explanation for the visible order of things than one which proceeds upon the assumption that there is nothing else. In this respect we may liken it to the hypothesis of atoms, or that of an ethereal medium, for neither of which we have the direct evidence of our senses, both of which have nevertheless been adopted as affording the best explanations of the phenomena of the visible universe. (74)

The Unseen Universe thus upholds speculative fiction over rigid statistical materialism. Its defense of analogy holds not only for its own invisible worlds, but also for any number of authoritative scientific hypotheses that are, at heart, conjectural.

Advocating imaginary science, Stewart and Tait fault the rigid authority of laws that "appear at first sight to hold exactly, or which, in other words, have the appearance of absolute truths" (98). They instead argue that "the laws of energy are rather generalisations. . . . There would be no permanent confusion of thought introduced if these laws should be found not to hold, or to hold in a different way, in the unseen universe" (210). This protest against homogenizing scientific law is indebted to the "counterfactual speculations"[42] of their friend and fellow physicist James Clerk Maxwell (1831–1879)—and especially to Maxwell's "Demon." In an 1867 letter to Tait, Maxwell first mentions this "very small BUT lively being" as a thought experiment devised to "show that the 2nd Law of Thermodynamics has only a statistical certainty."[43] Working against entropy, Maxwell's being serves as an intelligent "valve,"[44] sorting the fast- and slow-moving molecules of a gas through a small trapdoor into separate hot and cold chambers, and thus defying the tendency toward uniform temperature that would otherwise result from thermodynamic equilibrium. According to critic Daniel Brown, Maxwell likely viewed *The Unseen Universe* as "an intemperate version of his own scientific efforts to understand the unseen by analogy with the seen."[45] Nonetheless, as Brown notes, his demon shares with Stewart and Tait an emphasis on anomaly and theistic free will in the face of reductive scientific determinism.[46] *The Unseen Universe* repeatedly stresses these "very slight deviations from exactitude," which we will inevitably discover in natural law "as time passes on, and our instruments become more delicate" (86).

Continually subject to such fine deviations, atoms and molecules arguably serve as one such statistical fiction in Victorian physics. For Stewart and Tait, atoms serve as prominent figures of natural theology: their "physical properties . . . form the alphabet which is put into our hands by God," as accommodating aids for reading "that Great Book which we call the universe" (238). As objects of divine design, molecules display what both Maxwell and John Herschel deem "the essential character of a manufactured article."[47] Their "exact equality" and material consistency "preclude . . . the idea of [their] being eternal and self-existent"[48] and presumably reveal their creation from nothing, albeit through religious rather than mathematical generalization. More than a divine mystery, however, the single molecule also remains a profoundly unknown—and unpredictable—site in existing scientific law. Tyndall recognizes them as "mental image[s] of the ultra-sensible,"[49] and, as Maxwell warns, physical experiments can only address "millions of molecules"; we therefore "cannot . . . ascertain the actual motion of any one of these molecules" from "statistical information."[50] The individual particle is thus a fertile realm of potential anomaly, much like *The Unseen Universe*'s own cosmic breaks and divergences. For, while the atom's physical uniformity would apparently gesture to God—or at least to "the point at which Science must stop"[51]—its *departure* from the aggregate poses a different scientific challenge.

Yet beyond individual or multiple atoms, the ether forms the most powerful speculative frontier in *The Unseen Universe,* as a "continuous medium"[52] that invisibly unites its recursive worlds while also allying Christian and scientific worldviews. Widely accepted in late-Victorian physics, this plenum is composed of material "of a different and higher order than that of ordinary matter."[53] *The Unseen Universe* defines it as a viscous, semifluid substance, consisting of both closed vortex rings and vortex filaments[54] that transmit light, heat, and all other energy through their subtle "vibratory motion" (149). For Stewart and Tait in particular, the ether offers a powerful symbolic solution to the organization of matter, the structure of multiple worlds, and the mechanics of spiritualized energy transmission. Influenced by the philosopher Thomas Young, they craft this system as a recessive scale of vortex rings in the image of a divine Great Chain of Being.[55] In addition to fluid models of bubbles, Stewart and Tait compare its structure both to concentric smoke circles and to an infinite chain of nested knots, represented by the "symbolic monogram"[56] of a trefoil knot on the title page of their first two editions. These recursive worlds are ordered according to increasingly rarefied levels of substance, ranked from lower to higher matter and, respectively, from lesser to greater energy (219). In sum, they form "an infinite series of Universes," together containing "an infinite store of energy" (222). This transfer and distribution of energy crucially exploits entropy in its distinctive model of the ether as an imperfect fluid subject to minute amounts of friction. *The Unseen Universe* celebrates this fractional process of loss and absorption, through which all of "the available energy of the visible universe will ultimately be appropriated by the ether" (157) and reproduced in eternity.

Tellingly, Stewart and Tait stray from scientific consensus in their portrait of the ether. Treating this substance not as a perfect plenum but as a leaky fluid subject to waste and transformation, our authors imagine an imperfectly redemptive world, populated by lively demons, disorderly atoms, and invisible deviations. Baroque and eccentric, these speculative fictions challenge the deterministic rigidity of materialism, fueling a contest of validity that indelibly marks both established Victorian science and its imaginary complement in *The Unseen Universe.* In the resulting climate of debate, the ether wielded the authority of an active experimental site where "a whole order of forces was waiting to be discovered."[57] As we shall see, *The Unseen Universe* skillfully exploited these undiscovered forces, in a narrative of energy transfer that inspired many emulators.

Afterlives of the Unseen Universe*: Blavatsky and James*

In his review of *The Unseen Universe,* Clifford fears for its future influence: "The alleged possibilities which [the authors] have opened up will not be used only in the way which they themselves could wish. Put ever so innocent a breadknife into the hands of a maniac or a murderer and it will be not one whit less dangerous because it was never intended to cut flesh" (781). Clifford believes this work will embolden spiritualists and fuel primitive superstition—if not worse. Directly addressing Stewart and Tait, he warns, "That which you keep in your heart, my brothers, is the slender remnant of a system which has made its red mark on history, and still lives to threaten mankind" (793). Shaped by Stewart and Tait's Christianized physics, Blavatsky's Theosophical Society certainly supports Clifford's premonitions, although hardly in their most luridly violent form. Moreover, *The Unseen Universe* remains the surprisingly forgotten premise for ongoing debates in philosophy and the emerging discipline of psychology, including claims made by James both in an initial 1875 review and in his later essay, "The Will to Believe." For James and Blavatsky, as for other thinkers who succeeded them, this thermodynamic fantasy continued to spur controversy as a limit case for scientific belief and spiritual persuasion.

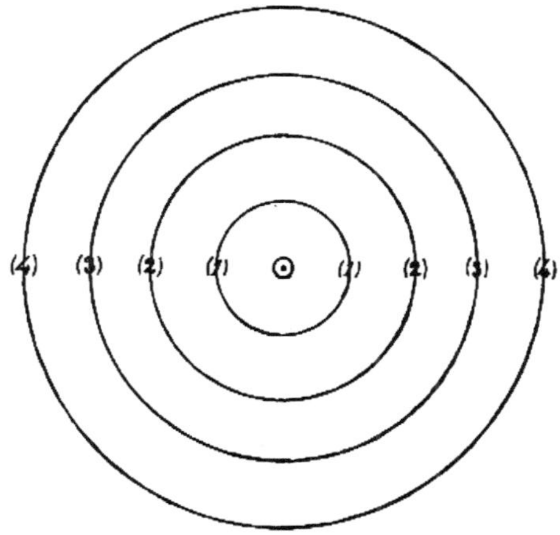

Fig. 13.1. Concentric circles, illustration in [Balfour Stewart and Peter G. Tait,] *The Unseen Universe, or, Physical Speculations on a Future State,* 2nd ed. (London: Macmillan, 1875), 171. (Public domain Google e-book.)

Untroubled by criticism of Stewart and Tait's methods, late Victorian mystics and occultists avidly cited *The Unseen Universe* as authoritative evidence. Especially vocal among these emulators was Blavatsky, who repeatedly invokes this text in her account of theosophy as a system of occult knowledge based on the ether's transmission of invisible spiritual forces and astral forms.[58] For, just as Stewart and Tait use the ether to challenge orthodox science, so too does Blavatsky question their privileged access to the topic in specialized professional circles. Brandishing *The Unseen Universe*'s rhetorical breadknife, she asks, "If scientists [may] indulge in . . . speculations" on ethereal energy transfer, "why should occultists and spiritualists be refused the same privilege?"[59] Although Blavatsky quotes Stewart and Tait with appreciation, she also argues that occultists have long pursued such inquiry beyond the mantle of scientific authority. In *Isis Unveiled,* she notes, "What the intelligent explorers of the *Unseen Universe* speculate upon, [is] to the masters of hermetic philosophy familiar science. To them ether [is] not merely a bridge connecting the seen and unseen sides of the universe," but a "road" whose "mysterious gates" they have already unlocked and entered (*Isis* 189). Combining spiritualist practices with a "synthesis of Western esotericism and . . . primarily Buddhism and Hinduism,"[60] Blavatsky's Theosophical Society promises the acquirement of secret wisdom allied with "cabalist, neo-Platonic, and Hermetic" forms of thought.[61] Even so, her writings most closely ally the movement with a vision of latent and universal energy, based on the validating language of thermodynamics and scientific law. In her *Key to Theosophy* (1889), Blavatsky advises all adepts "to

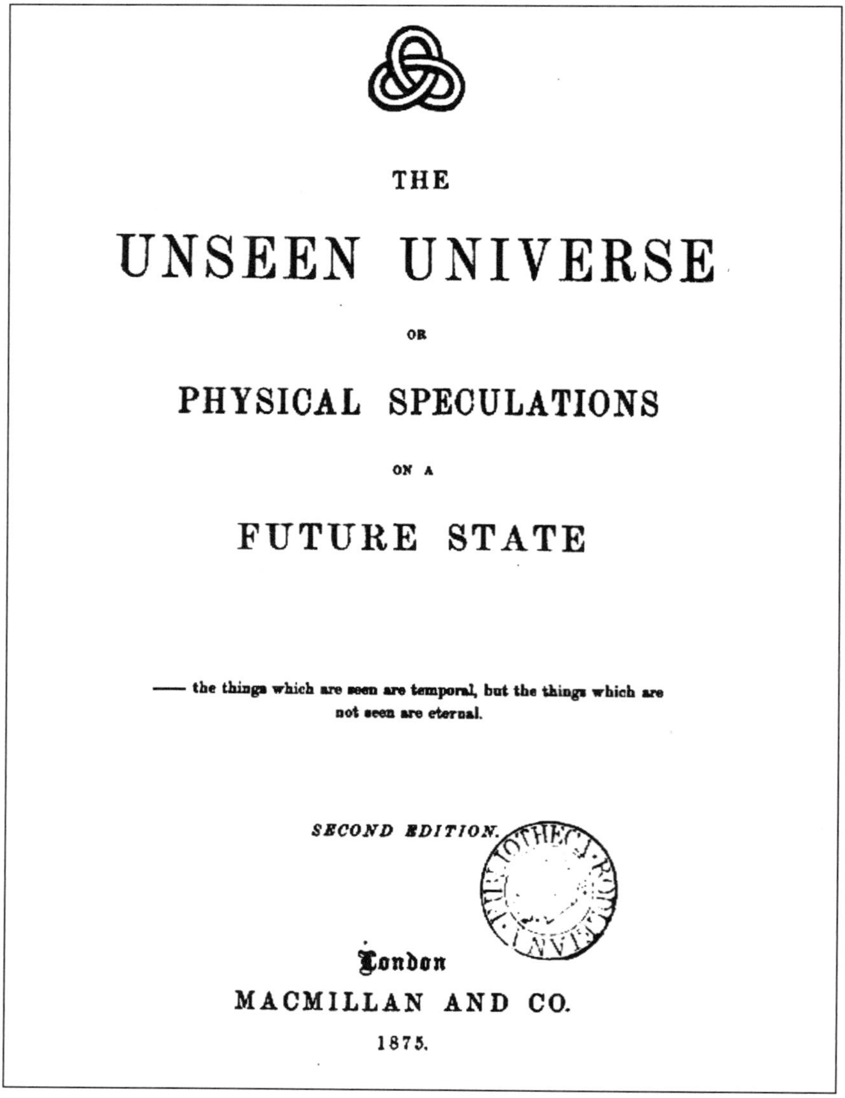

THE

UNSEEN UNIVERSE

OR

PHYSICAL SPECULATIONS

ON A

FUTURE STATE

— the things which are seen are temporal, but the things which are not seen are eternal.

SECOND EDITION.

London
MACMILLAN AND CO.
1875.

Fig. 13.2. Trefoil knot from the title page, illustration in [Balfour Stewart and Peter G. Tait,] *The Unseen Universe, or, Physical Speculations on a Future State*, 2nd ed. (London: Macmillan, 1875), iii. (Public domain Google e-book.)

investigate the hidden mysteries of Nature under every aspect possible, and the psychic and spiritual powers latent in man especially."[62] These frequent allusions to spiritual force seek to legitimate theosophy as an authoritative form of science, grounded in the physical and metaphysical study of nature.

Through her mythology of traveling astral forms and thought-impulses, Blavatsky embraces Stewart and Tait's redemptive fantasy of the universe's latent and ever-abundant energy, as Bruce Clarke has extensively shown.[63] She describes astral souls as "centre[s] and engine[s] of force, fed from the universal supply of force, and moved by the same general laws which pervade all nature and produce all cosmical phenomena" (*Isis* 197–98). Once this astral "*perisprit*" leaves the confines of its living human body, it moves through the ether, "alight[ing] at the threshold of the mysterious 'bourne'" between seen and unseen worlds, where it "can look but through a chink" (*Isis* 159). Like the revelatory breaks of *The Unseen Universe,* this "chink" marks a site of spiritual transfer and conversion, promising greater mystical vision. In *Isis Unveiled,* the ether transmits these astral souls and influences as "thought-impulse[s]" carried "through the ever-interchanging currents of energy between the two worlds, the visible and the invisible, from one succeeding age to another, until it affects a large portion of mankind" (*Isis* 181). This ambitious vision of force and influence dominates *Isis Unveiled* and *The Unseen Universe.* For, as Clarke recognizes, both texts view the ether as an invisible and universal "medium of information storage and retrieval"—a "cosmic data bank"[64] that captures every thought and every action as transferred and recuperated power. These totalizing fictions of energy model the greater logic of *The Unseen Universe,* as a form of imaginary science supporting expansive symbolic systems of faith, persuasion, and consolation.

James's review of Stewart and Tait's text responds, above all, to this expansiveness. His essay for the *Nation* marvels at their treatise's sheer "vastness of scale" (*Essays* 293)—its grandiose reshaping of thermodynamics into a cosmic geography of belief. Like Blavatsky, he views this text as inseparable from modern debate surrounding scientific authority and imaginative speculation. In fact, James notes, *The Unseen Universe* is entirely consistent with the "vast theories" of other established scientists, which remain as "unverified to-day . . . as any of the theosophies of the past" (*Essays* 115). Offering a nod to mystics and spiritualists, he argues that we must treat all grand speculators equally, leaving no room for scientific exception in our current "age of synthesis" (*Essays* 115). As

James therefore urges, grand unified theory should "be distinctly recognized for what it is—the mood of Faith, not Science" (*Essays* 115). James certainly shows ambivalence toward Stewart and Tait's use of traditional religious rhetoric, which he faults as "precisely identical with those of the most primitive, 'unscientific,' and short-winded natural theologian" (*Essays* 293). Discounting the emotional effects of analogy and accommodation, he treats these rhetorical forms as serving "apparently no function but to keep up logical appearances" (*Essays* 293). Nonetheless, the philosopher commends the magnitude of Stewart and Tait's greater vision. "It is," he reflects,

> only the incomparable superiority of the modern scientific imagination on its *quantitative* side that . . . give[s] our author[s'] speculations a different aspect . . . [Stewart and Tait] may enjoy the credit of having attacked the problem of natural theology (and solved it in [their] way) on a scale not unworthy of the grandeur of the theme's dimensions. (*Essays* 293)[65]

Indeed, James suggests that, as argument, the deep structure of *The Unseen Universe* lies in this massive spatial and quantitative scale, which uses the distinctive terms and dimensions of modern physics to model a new affective landscape of modern belief. For James, the *Unseen Universe* does *not* establish a satisfactory continuity between mechanical reasoning about facts and teleological reasoning about final causes. However, it still "widens the data and horizon which teleology receives from science" (*Essays* 294). For those reasons alone, *The Unseen Universe* should inspire trust in readers—if that trust makes a difference. Or, to use James's own words, "We for our part not only hold that such an act of trust is licit, but we think, furthermore, that any one *to whom it makes a practical difference* (whether of motive to action on or of mental peace) is in duty bound to make it" (*Essays* 293–94). Here James promotes a stance of speculative receptiveness toward "trust" and allied forms of belief—as impulses crucially spurred by human interest, emotion, and, even, duty.

Critics have subsequently questioned James's use of the term "duty" to describe this act of belief. In a later debate with the philosopher Chauncey Wright, James sought to retract the term for one reflecting a more "prudential," rather than simply "moral," obligation to believe.[66] Although James eventually replaced the term "duty" with "will," biographer Ralph Perry argues that "his favourite replacement for it was 'The Right to Believe,'"[67] presumably because "there was . . . an equal right *not* to believe."[68] Philosophers continue to debate how free or constrained James's notion of belief is—and whether we should view it as a duty or a "liberty" that "we are *free* to choose."[69] Notably, James emphasizes the context and consequences of such belief—that its pursuit might make "a practical difference." In this respect, we are "duty bound" to believe *if* it benefits us. As critic James Wernham notes, James here ultimately supports an "ought to believe" or "foolish-not-to-believe doctrine":[70] the double negative (we may not *not* believe) recalls Stewart and Tait's own claims that they "cannot deny the possibility of a future state" (6). Like the authors of *The Unknown Universe*, James advocates for openness in the face of uncertainty and constraint.

Upholding receptivity and intellectual possibility, James's review shaped his later renowned essay "The Will to Believe."[71] This lengthier essay argues "*in* justification *of* faith, [as] a defense of our right willfully to adopt a believing attitude in religious matters, in spite of the fact that merely logical intellect may not have been coerced."[72] Here, too, James supports the opportunity of belief, despite incomplete or inadequate proof. He urges us to resist our scientific fear of credulous belief—"the awful risk of believing lies" (*Will* 18)—an impulse that both Clifford and Tyndall openly state as motivating their empirical approaches. James even goes so far as to suggest that "science has organized this nervousness into a regular *technique*, her so-called method of verification," which cares not "for truth by itself at all," but "only truth as technically verified" (*Will* 21). He questions not only the defensive, technical aspects of scientific method but also—like Blavatsky—the double standard enjoyed by scientists free to speculate at will: these individuals "yield to the pleasures of taking for true what they happen vividly to conceive as possible" (*Essays* 115). Instead, James proposes an affective stance of openness and speculation that supports belief in the unseen, the uncertain, and the unknown. In the face of modern science and its battery of empirical limits and techniques, he dwells on questions of readerly disposition. Stressing the human costs and benefits of belief, James urges a receptivity that recalls Stewart and Tait's accommodation of "honest doubters" (202), as well as the flexible, speculative form of their treatise itself.

James's response to the rallying cry "believe nothing" may seem an obscure chapter in the mixed destinies of late Victorian natural theology and thermodynamics. Yet, as refracted through both his lens and that of occult practitioners, *The Unseen Universe* highlights still-urgent questions surrounding belief, persuasion, imagination, and, above all, speculation. As a form of modern spiritual and scientific consolation, speculative openness is this text's—and, indeed, also James's—enduring stance toward a closed and finite world. For, absent truly positive evidence or revelation, Stewart and Tait suggest the best approach may simply be to maintain imaginative openness in the face of physical and intellectual limits. In *The Unseen Universe*, this speculative flexibility overcomes barriers through the sheer persuasive force of thermodynamic analogy in all of its vastness. Recasting limitation as a field of fanciful possibility, Stewart and Tait explore a baroque range of scientific fictions, theistic models, and rhetorical devices, ranging from atoms and the ether to bubble-forms and heat-engines. Nonetheless, energy and entropy remain its greatest source of imaginative transformation, spurring both this text's central dilemma—the end of the world—and its symbolic solution—alternate realms. Sustaining the leaky chain of cosmic conversion and redemption, this energy not only assumes the speculative structure of belief; it also tests the rival claims of science and belief themselves as a site of rhetorical convergence that intensely questions the limits of knowledge and the legitimacy of evidence. For, as James observes in "The Will to Believe," in many cases of established scientific theory, the distinction between hard fact and speculative faith is nothing if not fluid: "Our faith is faith in some one else's faith, and in the greatest matters this is most the case" (*Will* 9). Beyond its traces in mysticism and theosophy, *The Unseen Universe*'s most formative legacy lies in this ever-urgent debate surrounding the unstable and potentially fictive basis of scientific truth claims. As James, Blavatsky, and our two authors show, both orthodox science and *The Unseen Universe* share a deep mythic and speculative core, as immersive technical and symbolic representations designed—like natural theology—to evoke belief. Through Stewart and Tait's inheritors, this expansive spiritual vision still inspires and attends us today, in the fantastic abstractions of both imaginary science and its modern symbolic complement, science fiction.

Notes

I would like to thank Shalyn Claggett, Lara Karpenko, and Daniel Youd for their invaluable help in revising this chapter.

1. For an especially nuanced account of the first and second laws of thermodynamics, and of the corresponding literary imagination, see Barri J. Gold, *ThermoPoetics: Energy in Victorian Literature and Science* (Cambridge: MIT Press, 2010).

2. George Levine, *Darwin and the Novelists: Patterns of Science in Victorian Fiction* (Cambridge: Harvard University Press, 1988), 156; Gillian Beer, *Open Fields: Science in Cultural Encounter* (Oxford: Clarendon Press, 1996), 262.

3. H. G. Wells, *The Time Machine* (Harmondsworth: Penguin, 2005), 82.

4. W[illiam] K[ingdon] Clifford, "The Unseen Universe," *Fortnightly Review* 17, no. 102 (June 1875): 792. Hereafter cited parenthetically.

5. Cargill Gilston Knott, *Life and Scientific Work of Peter Guthrie Tait: Supplementing the Two Volumes of Scientific Papers Published in 1898 and 1900* (Cambridge: Cambridge University Press, 1911), 236–37. Arthur Schuster, Stewart's successor at Owen's College, believed Tait "played a very secondary role in the writing of the *Unseen Universe*. The work was largely Stewart's." See Schuster, *Biographical Fragments* (London: Macmillan, 1932), 214–25, quoted in Janet Oppenheim, *The Other World: Spiritualism and Psychical Research in England, 1850–1914* (Cambridge: Cambridge University Press, 1985), 472.

6. [Balfour Stewart and Peter G. Tait,] *The Unseen Universe, or, Physical Speculations on a Future State*, 2nd ed. (London: Macmillan, 1875), xi.

7. Balfour Stewart and Peter G. Tait, *The Unseen Universe, or, Physical Speculations on a Future State* (1901; New York: Cosimo Classics, 2007), 5. Hereafter cited parenthetically.

8. Crosbie Smith, *The Science of Energy: A Cultural History of Energy Physics in Victorian Britain* (Chicago: University of Chicago Press, 1998), 254.

9. Jay Clayton, *Charles Dickens in Cyberspace: The Afterlife of the Nineteenth Century in Postmodern Culture* (Oxford: Oxford University Press, 2006), 8.

10. Greg Myers, "Nineteenth-Century Popularizations of Thermodynamics and the Rhetoric of Social Prophecy," *Victorian Studies* 29, no. 1 (1985): 50; Robert Hughes Kargon, *Science in Victorian Manchester: Enterprise and Expertise* (Baltimore: Johns Hopkins University Press, 1977), 214–24.

11. Iwan Rhys Morus, *When Physics Became King* (Chicago: University Chicago Press, 2005), 140.

12. Knott, *Life and Scientific Work*, 236; John Tyndall, *Fragments of Science: A Series of Detached Essays, Addresses, and Reviews*, vol. 2 (New York: D. Appleton, 1898), 197.

13. Tyndall, *Fragments of Science*, 190.

14. Graeme Gooday, "Sunspots, Weather, and the Unseen Universe: Balfour Stewart's Anti-materialist Representations of 'Energy' in British Periodicals," in *Science Serialized: Representations of the Sciences in Nineteenth-Century Periodicals*, ed. Geoffrey N. Cantor and Sally Shuttleworth (Cambridge: MIT Press, 2004), 112.

15. P. M. Heimann, "The 'Unseen Universe': Physics and the Philosophy of Nature in Victorian Britain," *British Journal for the History of Science* 6, no. 1 (1972): 73.

16. Jonathan R. Topham, "Science, Natural Theology, and the Practice of Christian Piety in Early-Nineteenth-Century Religious Magazines," in Cantor and Shuttleworth, *Science Serialized*, 38. Topham cautions that the "discourse of design" or "argument from design" should not be treated as synonymous with natural theology but rather as only one outgrowth of this religious and rhetorical tradition (38).

17. Bernard Lightman, *The Origins of Agnosticism: Victorian Unbelief and the Limits of Knowledge* (Baltimore: Johns Hopkins University Press, 1987), 152–53.

18. John Hedley Brooke and R. Hookyaas, *New Interactions between Theology and Natural Science* (Milton Keynes: Open University Press, 1974), 42.

19. Gold, *ThermoPoetics*, 71. On the relation of *The Unseen Universe* to "grand unified" theories of energy physics, see 90–91, 126–28.

20. Ibid., 57–58. Gold compares the development of thermodynamic concepts to elegiac processes of consolation in Alfred Tennyson's *In Memoriam* (1850).

21. Bruce Clarke, "Allegories of Victorian Thermodynamics," *Configurations* 4, no. 1 (1996): 14.

22. Helge S. Kragh, *Entropic Creation: Religious Contexts of Thermodynamics and Cosmology* (Aldershot: Ashgate, 2008), 47.

23. As defined by Stewart and Tait, this pseudoscientific principle allies cause-and-effect logic with both the law of conservation and the teleological demands of natural theology. Widely faulted for its vagueness and imprecision, the "Law of Continuity" expresses the consistency, uniformity, and comprehensibility of the universe's combined visible and invisible worlds (271). See also Kragh, *Entropic Creation*, 64.

24. Postmodern models of the multiverse differ importantly from Stewart and Tait's, however, as they restrict any physical interchange between their parallel worlds. On multiple bubble-worlds, see Brian Greene, *The Elegant Universe: Superstrings, Hidden Dimensions, and the Quest for the Ultimate Theory* (New York: W. W. Norton, 2010).

25. Michel Serres, *Hermes: Literature, Science, Philosophy*, trans. Josué V. Harari and David F. Bell (Baltimore: Johns Hopkins University Press, 1982), 116.

26. Ibid., 72, 71–72.

27. On mechanical and thermodynamic accounts of life, see David Channell, *The Vital Machine: A Study of Technology and Organic Life* (Oxford: Oxford University Press, 1991), 81–83; Tamara Ketabgian, *The Lives of Machines: The Industrial Imaginary in Victorian Literature and Culture* (Ann Arbor: University of Michigan Press, 2011), 121–24; and Serres, *Hermes*, 71–83.

28. On Carnot's engine, see Stewart and Tait, *Unseen Universe*, 117–21.

29. William Paley, *Natural Theology* (London: J. Faulder, 1809), 12. Hereafter cited parenthetically.

30. Tamara Ketabgian, "Prosthetic Divinity: Babbage's Engine, Spiritual Intelligence, and the Senses," Victorian Review 35, no. 2 (2009): 33–34.

31. Arnold Huijgen, *Divine Accommodation in John Calvin's Theology: Analysis and Assessment* (Göttingen: Vandenhoeck and Ruprecht, 2011), 14.

32. Edward A. Dowey, *The Knowledge of God in Calvin's Theology* (New York: Columbia University Press, 1965), 10.

33. Ketabgian, "Prosthetic Divinity," 33.

34. As Gooday notes, this explanation of "delicacy" originated in an earlier journal article coauthored by Stewart and J. Norman Lockyer, "The Sun as a Type of the Material Universe," part 1, *Macmillan's Magazine* 18 (1868): 246–57. See Gooday, "Sunspots," 113.

35. Oppenheim, *Other World*, 336–37; Balfour Stewart, "Address IV, April 23rd, 1887," in *Presidential Addresses to the Society for Psychical Research, 1882–1911* (Glasgow: Robert Maclehose, 1912), 28. Before the founding of the Society for Psychical Research in 1881, Stewart analyzed chemist William Crookes's experiments on the production of "psychic force" in séances by medium Daniel Dunglas Home in 1871. Stewart concluded that Crookes's findings were subjective and inconclusive. On the "evidential crisis of orthodox science" that followed, see Peter Lamont, "Spiritualism and a Mid-Victorian Crisis of Evidence," *Historical Journal* 47 (2004): 911–19; and Balfour Stewart, "Mr Crookes on the 'Psychic' Force," *Nature*, July 27, 1871, 237.

36. Richard Noakes, "The Sciences of Spiritualism in Victorian Britain," in *The Ashgate Research Companion to Nineteenth-Century Spiritualism and the Occult*, ed. Tatiana Kontou and Sarah Willburn (Burlington, VT: Ashgate, 2012), 41–42. On the failed séance, see also John Tyndall, "Science and the Spirits," *Reader* 4 (1864): 725–26; and [William Henry Harrison], "Professor Tyndall at a Spirit Circle," *Spiritualist* 1 (1871): 156–57.

37. Humphrey Palmer, *Analogy: A Study of Qualification and Argument in Theology* (New York: St. Martin's Press, 1973), 15–16.

38. William James, *Essays, Comments, and Reviews* (Cambridge: Harvard University Press, 1987), 292. Hereafter cited parenthetically as "*Essays*."

39. Emphasis mine.

40. Emphasis mine.

41. Istvan Csicsery-Ronay, *The Seven Beauties of Science Fiction* (Middleton, CT: Wesleyan University Press, 2008), 111.

42. Daniel Brown, *Poetry of Victorian Scientists: Style, Science, and Nonsense* (Cambridge: Cambridge University Press, 2013), 254.

43. Qtd. in Knott, *Life and Scientific Work*, 214–15.

44. Ibid., 215.

45. Brown, *Poetry of Victorian Scientists*, 254.

46. Ibid.

47. James Clerk Maxwell, *The Scientific Papers*, vol. 2 (Cambridge: Cambridge University Press, 1890), 376.

48. Ibid.

49. Tyndall, *Fragments of Science*, 103.

50. Maxwell, *The Scientific Papers*, 374.

51. Ibid., 376.

52. Daniel M Siegel, "Thomson, Maxwell, and the Universal Ether in Victorian Physics," in *Conceptions of Ether: Studies in the History of Ether Theories, 1740–1900*, ed. G. N. Cantor and M. J. S. Hodge (Cambridge: Cambridge University Press, 1981), 258.

53. Richard Noakes, "Ethers, Religion and Politics in Late-Victorian Physics: Beyond the Wynne Thesis," *History of Science* 43 (2005): 422.

54. Siegel, "Thomson," 258.

55. G. N. Cantor, "The Theological Significance of Ethers," in Cantor and Hodge, *Conceptions of Ether*, 140.

56. Knott, *Life and Scientific Work*, 106.

57. Morus, *When Physics Became King*, 174.

58. Clarke, "Allegories," 87–88.

59. Helena Petrovna Blavatsky, *Isis Unveiled: A Master-Key to the Mysteries of Ancient and Modern Science and Theology*, 6th ed., vol. 1 (New York: J. W. Bouton, 1891), 185. Hereafter cited parenthetically as "*Isis*."

60. Mark Morrisson, "The Periodical Culture of the Occult Revival: Esoteric Wisdom, Modernity, and Counter-public Spheres," *Journal of Modern Literature* 31, no. 2 (2008): 7.

61. Oppenheim, *Other World*, 163.

62. Helena Petrovna Blavatsky, *The Key to Theosophy: Being a Clear Exposition, in the Form of Question and Answer, of the Ethics, Science and Philosophy for the study of which the Theosophical Society has been founded* (London: Theosophical Publishing Company, 1889), 39.

63. See Clarke, "Allegories," 87–88.

64. Ibid., 88.

65. I have revised this passage to reflect *The Unseen Universe*'s plural authorship, of which James was presumably ignorant upon the review's publication in May 1875.

66. James C. S. Wernham, *James's Will-to-Believe Doctrine: A Heretical View* (Kingston: McGill-Queen's University Press, 1987), 13–15; Edward H. Madden, *Chauncey Wright*

and the Foundations of Pragmatism (Seattle: University of Washington Press, 1963), 43–45.

67. Ralph B. Perry, *The Thought and Character of William James*, vol. 2 (Boston: Little, Brown, 1935), 244. According to Perry, James later regretted his original title for the essay (244–45).

68. Wernham, *Doctrine*, 6.

69. Dickinson S. Miller, "James's Doctrine of 'The Right to Believe,'" *Philosophical Review* 51, no. 6 (1942): 541.

70. Wernham, *Doctrine*, 15. Wernham's phrase alludes to Wright's claim that during their debate James intended to argue "it is foolish not to believe, or try to believe, if one is the happier for believing" (14).

71. On parallelism in the language of James's 1875 review and his later 1896 essay, see Wernham, *Doctrine*, 16; and Madden, *Chauncey Wright*, 43.

72. William James, *The Will to Believe and Other Essays in Popular Philosophy* (New York: Longmans, Green, 1912), 1–2. Hereafter cited parenthetically as "*Will*."

Contributors

Editors

Lara Karpenko is Associate Professor of English at Carroll University. Her research interests include Victorian literature and culture, gender studies, readership history, and the novel. She, along with Lauri Dietz (DePaul University) is the coeditor of a special issue of *Nineteenth-Century Gender Studies* focused on effective pedagogies, and her recent work has appeared in the *Victorian Review*, *Nineteenth-Century Studies*, and the *Victorians Institute Journal*.

Shalyn Claggett is Associate Professor of English at Mississippi State University, specializing in nineteenth-century British literature and science. Her current book project, "Equal Natures: The Science of Character in Victorian Women's Writing," examines how women authors engaged with the popular science of phrenology. Her work has appeared in such journals as *Victorian Literature and Culture*, *SEL*, *Victorians Institute Journal*, and *Prose Studies*.

Contributors (in order of appearance)

Gillian Beer is King Edward VII Professor Emeritus at the University of Cambridge and former president of Clare Hall College. She is a Fellow of the British Academy and of the Royal Society of Literature. She is also a Foreign Honorary Member of the American Academy of Arts and Sciences and an Honorary International Member of the American Philosophical Society. She was made a Dame in 1998. Among her books are *Darwin's Plots: Evolutionary Narrative in Darwin* (2009), *George Eliot and Nineteenth-Century Fiction* (1983), *Open Fields: Science in Cultural Encounter* (1996), and *Alice in Space: The Sideways Victorian World of Lewis Carroll* (2016).

Lynn Voskuil is Associate Professor of English at the University of Houston, where she teaches Victorian studies and empire studies. She is currently completing "Horticulture and Imperialism: The Garden Spaces of the British Empire, 1789–1914." Supported by an NEH fellowship, this book project explores nineteenth-century Britain's obsession with tropical plants and horticulture. Her essays have appeared in such journals as *Victorian Studies, ELH, Feminist Studies,* the *Journal of British Studies,* and *Studies in Romanticism,* and in a variety of collections published by Cambridge, Oxford, and Blackwell. She is the author of *Acting Naturally: Victorian Theatricality and Authenticity* (University of Virginia Press, 2004) and the editor of *Nineteenth-Century Energies: Literature, Technology, Culture* (Routledge).

Meegan Kennedy is Associate Professor of English at Florida State University and a core faculty member with the History and Philosophy of Science Program. Her research interests include Victorian literature and culture, the British novel, Victorian science and medicine, and periodical studies. She is particularly interested in nineteenth-century theories of visuality and representation. She has published *Revising the Clinic: Vision and Representation in Victorian Medical Narrative and the Novel* (Ohio State University Press, 2010), which examines the interplay between medical case histories and British novels from the eighteenth century to the age of Freud. Her current book project, "Beautiful Mechanism," examines Victorians' romance with the microscope and its analogue, the eye.

Narin Hassan is Associate Professor in the School of Literature, Communication, and Culture at the Georgia Institute of Technology, where she teaches courses on Victorian literature and culture, postcolonial and gender studies, and the history and culture of medicine. Her book *Diagnosing Empire: Women, Medical Knowledge, and Colonial Mobility* was published by Ashgate in 2011, and her most recent articles appear in *Women's Studies Quarterly, Mosaic,* and *Nineteenth-Century Gender Studies.* She is currently researching material on gender, landscape, and empire for a book tentatively titled "Troubling Natures: Nineteenth-Century Cultures of Botany and Empire."

Elizabeth Chang is Associate Professor of English at the University of Missouri, where she specializes in nineteenth-century British literature, particularly literature of the nineteenth-century British Empire. She is the author of *Britain's Chinese Eye: Literature, Empire and Aesthetics in Nineteenth-Century Britain* (Stanford University Press, 2010) and the editor of the collection *British Travel Writing from China, 1793–1901* (Pickering and Chatto, 2009).

Danielle Coriale is Assistant Professor of English at the University of South Carolina. Her research explores the relationship between literature and science in nineteenth-century Britain. Her work has appeared in *Nineteenth-Century Contexts, Victorian Review, Nineteenth-Century Literature,* and *SEL.* She is currently completing a book manuscript entitled "Zoology and Victorian Literature: Reading Humans and Other Animals."

James Emmott is a Lecturer in Victorian Studies in the Department of English at Queen Mary University of London. His work has appeared in *Victorian Studies* and the *Journal of Victorian Culture,* and he is currently writing a book about the techniques and technologies of recording and reading the voice and face in nineteenth- and early twentieth-century science and culture.

Suzanne Raitt is Chancellor Professor of English at the College of William & Mary. Her books include *May Sinclair: A Modern Victorian* (Oxford University Press, 2000), *Vita and Virginia: The Work and Friendship of V. Sackville-West and Virginia Woolf* (Oxford University Press, 1993), and *Virginia Woolf's "To the Lighthouse"* (St Martin's, 1990). She also coedited a collection of essays with Trudi Tate called *Women's Fiction and the Great War* (Oxford University Press, 1997), and in 1995 she published an edited collection of essays on lesbian criticism, *Volcanoes and Pearl Divers* (Onlywomen Press). Her editions include a Norton Critical Edition of

Virginia Woolf's *Jacob's Room* in 2007, Katherine Mansfield's *Something Childish and Other Stories* for Penguin in 1996, and Virginia Woolf's *Night and Day* for Oxford World's Classics in 1992. She is currently working on a scholarly edition of Virginia Woolf's *Orlando*, coedited with Ian Blyth, for Cambridge University Press; and completing a series of articles on waste and efficiency in late nineteenth- and early twentieth-century British culture.

Barri J. Gold is Professor and Chair of English Language and Literature at Muhlenberg College. She earned a BS in physics from MIT and a PhD in English language and literature from the University of Chicago. Her book *ThermoPoetics: Energy in Victorian Literature and Science* (MIT Press, 2010) considers the work of such thinkers as Tennyson, Thomson, Joule, Maxwell, Dickens, Spencer, Bulwer-Lytton and Stoker as it explores the conversation between literature and physics that develops the laws, concepts, and implications of thermodynamics.

Sumangala Bhattacharya is Associate Professor of British Literature at Pitzer College in Claremont, California. She earned an AB in physics from Smith College and a PhD in English from the University of Southern California. Her published work has appeared in journals such as *Clio, Nineteenth-Century Contexts,* and *Nineteenth-Century Gender Studies* and focuses on issues surrounding gender and colonialism. Her current book project examines representations of hunger in the nineteenth century. She is also currently pursuing a JD at the Dale E. Fowler School of Law at Chapman University.

Anna Maria Jones is Associate Professor of English at the University of Central Florida, where she teaches Victorian literature, literary theory, and the history of the novel. Her book *Problem Novels: Victorian Fiction Theorizes the Sensational Self* was published in the Ohio State University Press's Victorian Critical Interventions Series in 2007. She is coeditor, with Rebecca N. Mitchell, of the forthcoming essay collection, *Drawing on the Victorians: The Palimpsest of Victorian and Neo-Victorian Graphic Texts* (Ohio University Press). Her recent articles have appeared in *Victorian Literature and Culture, European Romantic Review, Criticism,* and *Neo-Victorian Studies.* She is also a contributor to the *Cambridge Companion to Victorian Culture,* the *Blackwell Companion to Sensation Fiction,* and *BRANCH.*

L. Anne Delgado recently completed her PhD in English at Indiana University. Her dissertation, "Cosmic Plots: Occult Knowledge and Narratives of Belief," examines the impact that late nineteenth-century alternative movements including spiritualism, theosophy, and occultism had upon the emergence of new fictional modes. She is working on a project concerning late nineteenth-century mediumship and the cultural history of ectoplasm. She is currently a lecturer at Indiana University.

Tamara Ketabgian is Professor of English at Beloit College. She is the author of *The Lives of Machines: The Industrial Imaginary in Victorian Literature and Culture* (University of Michigan Press, 2011), in addition to various articles and book chapters on nineteenth-century literature and science. Her research has received fellowship support from the American Philosophical Society and the American Council of Learned Societies. She is currently working on a book project entitled "Contrivance: Faith, Persuasion, and Technology in Victorian Scientific and Literary Culture," which explores fantasies of technological design and spiritual intelligence from Charles Babbage to the present.

Index

Page numbers in italics refer to illustrations.